THE POLITICS OF
THE ENVIRONMENT

A SURVEY

T0227630

THE POLITICS OF
THE ENVIRONMENT

A SURVEY

FIRST EDITION

Edited by Chukwumerije Okereke

Routledge
Taylor & Francis Group

LONDON AND NEW YORK

First Edition 2007
Routledge
2 Park Square, Milton Park, Abingdon, Oxfordshire OX14 4RN
711 Third Avenue, New York, NY 10017

First issued in paperback 2015

Routledge is an imprint of the Taylor & Francis Group, an informa business

Development Editor: Cathy Hartley
Copy Editor and Proof-reader: Simon Chapman

Typeset in Times New Roman 10.5/12

The publishers make no representation, express or implied, with regard to the accuracy of the information contained in this book and cannot accept any legal responsibility for any errors or omissions that may take place.

Typeset by AJS Solutions, Huddersfield – Dundee

ISBN 13: 978-1-85743-756-0 (pbk)
ISBN 13: 978-1-85743-341-8 (hbk)

Contents

CONTENTS

Introduction

Over the last three or four decades, the environment has emerged as one of the major issues on the political agendas of governments at all scales—local, national and international. The rising profile of the environment in politics reflects the increasing perception that human beings have, through the patterns of their getting and spending, drastically affected the state of the planet and its ability to provide a support-base for many in the present and future generations. Some indeed have begun to express the opinion that environmental problems and specifically climate change may very well be the greatest challenge that humanity will face in the 21st century (King 2004; Vogler 2007: 435).

The growing concern that mankind may be facing a large-scale ecological crisis has led to an unprecedented level of activity by both state and non-state actors wishing to address various aspects of the environmental challenge. Politically, these activities implicate different notions of the role the state, the relationship between the public and the private, how international co-operation should be structured, who has authority to act, how the economy (national and global) should be organized, and the elemental moral question of how we should live and what values should guide individual and collective actions.

This reference volume aims to provide a state-of-the-art review of the politics of the environment with particular attention to the international dimension. The volume is intended to serve as a first stop information source for those wishing to gain a quick but informed overview of the trajectory and key themes that have shaped discussions on global environmental politics over the last four decades or so. The volume is divided into four parts: review essays; an A–Z glossary; and sections of select maps and statistics.

There are nine review essays written by experts and designed to provide authoritative accounts of developments and major debates involved in essential aspects of the international politics of the environment. The first essay, entitled 'Globalization: The Environment and Development Debate', by J. Timmons Roberts, characterizes the central challenge of environmental politics in terms of a 'tension between development goals and environmental protection'. The tension resides in the fact that while industrialization, increasing economic development and international trade are widely seen as vital conditions for improving human well-being, these activities on the other hand result in drastic changes to the environment with multiple immediate and long-term consequences. He provides an account of the attempts by states to negotiate this tension and highlights the existence of a condition of uneasy compromise where for the most part it appears that environmental concerns (especially those of the developing countries) are subordinated to the general quest for economic development.

The second essay, by Hugh Dyer, provides a brief but extremely thoughtful consideration of the link between the state and international relations on the one hand, and the theory and practice of environmental politics on the other. He indicates various reasons why the environment poses a unique challenge to traditional understandings of the state and patterns of international relations. These include, *inter alia*, the highly interdependent nature of the global ecosystem and the fact that environmental pollution does not respect political boundaries. After highlighting some of the ways in which the practice of environmental politics has pushed the frontiers of traditional political institutions, he leaves us to ponder whether existing structures and 'the traditional staples of the field—state, nation, international relations', despite their resilience, are indeed appropriate categories in thinking about effective solutions for global environmental problems.

While the state has been termed 'the preeminent institution with the requisite political authority and steering capacity to tackle ecological problems' (Barry and Eckersley 2005: ix), it is not by any means a static entity, nor an isolated player. Matthew Paterson, in the third essay, provides a highly nuanced account of the intricate connections between the state, the dynamics of the global economy and the strategies for responding to particular environmental challenges and indeed the pursuit of environmental sustainability in general. He does this through the elaboration of four key themes and questions around which debates about environmental politics and the international political economy are often framed. These include: the link between existing structures for managing the global economy and the environment; the power and role of multinational corporations in solving or exacerbating environmental problems; the relationship between economic growth and environmental sustainability; and the 'overall compatibility of a "free market" or "capitalist" economy with the principle of sustainability'. Of course, in keeping with the objective of the volume, Paterson does not attempt to argue in favour of or against any of the competing perspectives he identifies. The critical point he makes, though, is that these themes and questions constitute some of the most important ones in 'thinking politically about the environment and sustainability'.

The fourth essay, by Hannes R. Stephan and Fariborz Zelli, examines the role of international organizations in the management of global environmental problems. Their analysis concentrates not on issue-specific regimes *per se* but on globally operating entities that have an explicit environmental mandate, as well as those whose actions exert significant influence on environmental issues and policies. Following this criterion, they provide an elegant account of the history, structure and activities alongside a brief evaluation of notable organizations like the UN Environmental Programme (UNEP), the UN Development Programme (UNDP), the Global Environment Facility (GEF), the Commission for Sustainable Development (CSD) and the World Trade Organization (WTO). Stephan and Zelli note the critical role of the UN in the formation of these organizations and in the institutionalization of environmental governance more broadly. They consider that the overlapping and fragmented nature of international environmental organizations stems in part from the multifaceted nature of global environmental problems and in part from the complex makeup of the UN system through which most of these organizations came into being. The central challenge, they note, is therefore how to achieve greater co-ordination among these organizations in order to enhance efficiency without unduly compromising or sacrificing the interests of the environment in the process.

If international environmental organizations are overlapping and fragmented, the landscape of environmental movements is one of an even greater degree of multiplicity and diversity. In the fifth essay, Brian Doherty provides an expert mapping of the contours of the nearly 40 years of rapid growth and development of environmental movements and their roles in establishing the environment on the political agendas of governments. He notes that the degree of institutionalization that has taken place over the last four decades, and in particular the electoral fortune of many green parties in Western Europe, means that environmental groups could no longer be necessarily classified as non-state actors. At the same time, he notes that despite the 'globalization' of many environmental movements, both in terms of presence and focus there is little evidence to suggest that this process would eventually lead to the emergence of an umbrella global environmental movement. The reason, he writes, is that the institutionalization of environmental movements over the past four decades has occurred concurrently with an incredible degree of 'diversification in strategies and ideology', some of which reflect the different socio-political and economic contexts within which these groups operate. The more difficult and perhaps also more urgent question relates to the future role of environmental movements in both national and global contexts. On this point, Doherty suggests that while a number of environmental movements have now become more or less co-opted into the practices of governance, to the extent that they are no longer able to seriously challenge prevailing institutions and ideas, it would be an overstatement to suggest that environmental movements in their totality have lost steam and are unable to exert significant influence in the future trajectory of global environmental issues and policies. The space for influence, he says, is particularly broad for environmental movements operating in the global South.

The co-ordination of environmental decision-making and implementation processes in multi-level governance systems are the subject of the sixth essay, by David Benson and Andrew Jordan. Using the European Union as a case study, Benson and Jordan explore three different areas of political contestation that often characterize attempts by relatively autonomous political authorities to co-ordinate efforts in dealing with environmental problems. These include how to: (i) allocate responsibility for policy formulation, implementation and enforcement; (ii) structure interactions between the different levels of government involved; and (iii) deal with the issue of differences in the instruments and tools used to accomplish set objectives. They note that while the unique nature of the EU has afforded it opportunities for making notable progress in resolving these issues, there is still much evidence that demonstrates the intractable difficulties and complexities implicated in multi-level environmental governance.

In the seventh essay Maxwell Boykoff explores the relatively less rehearsed but none the less very significant link between the mass media and environmental politics. His account seeks to demonstrate that the mass media are not passive actors whose role is limited to that of merely reporting on the environment.

Rather, the mass media, through their choice of coverage and style of reporting, significantly influence the framing and development of environmental debates and policies. At the same time he notes that the mass media do not operate in a political vacuum, but are very much influenced by the wider political context within which they function. In other words, there is an intimate relationship between the nature of prevailing institutions and ideas and the ways in which the mass media choose to communicate the progress, challenges and disagreements that shape environmental politics.

The eighth essay, written by Tim Rayner and Chukwumerije Okereke, is the only one that focuses on a specific environmental issue—climate change. This issue has been singled out because climate change, as noted at the beginning of this introduction, has been described as the most complex environmental problem ever to be faced by humanity. The authors begin by identifying at least six features that make climate change a particularly difficult problem to manage. Among these are the long-term nature of the problem, the uncertainties surrounding the issue (which mean that policy-makers can never be sure that the action they have taken is adequate) and the fact that emissions are linked to such a wide array of human activities. After exploring the main dimensions of the challenge involved in building 'effective' climate governance at the global level, they conclude that although radical and urgent action is needed, the institutional hurdles in the way are formidable.

The final essay, by Chukwumerije Okereke, explores the link between ethics and international environmental politics. He attempts to demonstrate that environmental politics and public policy are far more embedded in deep normative and moral values than is often appreciated in mainstream political circles. He maps the influence of ethics in the development of global environmental awareness and policies and identifies democracy and effective participation as the two key challenges of institutions for global environmental governance. In conclusion, he predicts that ethical questions are likely to become increasingly central in the search for collective responses to environmental problems at the international level.

The second part of the book consists of an A–Z glossary section prepared by Chukwumerije Okereke. The entries are carefully selected to cover the essential concepts, themes, ideas and institutions that figure prominently in the discourse of global environmental politics. Entries are cross-referenced for ease of use.

The third and fourth sections provide a number of maps and statistical tables designed to aid the reader with key information on the distribution and pattern of use of some of the world's valuable natural resources.

The intention is that this volume should be a useful reference, teaching and training resource.

Chukwumerije Okereke
Tyndall Centre for Climate Change Research
University of East Anglia, UK
October 2007

Acknowledgements

The editor gratefully acknowledges the support of Routledge for this project. The editors, authors and publishers are grateful to the Cartographic Unit at the University of Southampton for drawing up the maps. They are also most grateful to the organizations which kindly gave permission for their data to be reproduced in the sections of maps and statistics.

The editor would also wish to thank the contributors to this volume from whom he learned so much in the editing process. Many thanks also to those who reviewed most of the chapters in this volume. Among these are John Vogler, Bettina Wittneben, Christopher Rootes, Mark Charlesworth, John Turnpenny and Chima Mordi.

The editor extends thanks to Simon Chapman for picking through the manuscripts and making very helpful editorial changes and to Cathy Hartley at Routledge for her able guidance, support and excellent editorial style.

The editor is indebted to his wife Boma Nkiruka for her understanding and support throughout the time it took to complete the project.

The Contributors

Chukwumerije Okereke is a Senior Research Associate at the Tyndall Centre for Climate Change and Research, Zuckerman Institute of Connective Environmental Research, School of Environmental Sciences, University of East Anglia. His main research interest is in theory and practice of global environmental governance. His latest work is *Global Justice and Neoliberal Environmental Governance: Ethics, Sustainable Development and International Co-operation* (Routledge 2008). His current research focuses on the interactions between corporate carbon strategies and future international climate policies.

David Benson is an ESRC Post Doctoral Fellow in CSERGE, which is located in the School of Environmental Sciences, University of East Anglia. He is particularly interested in the motivations or rationales of actors in allocating environmental tasks to specific levels in the European Union multi-level political system, and in federal approaches to EU integration more generally. He has also published on sustainability appraisal and environmental politics. He is currently collaborating in writing the third edition of *Politics and the Environment*, published by Routledge.

Maxwell T. Boykoff is a James Martin 21st Century Research Fellow at the University of Oxford. He holds a Ph.D. in Environmental Studies (with a parenthetical notation in Sociology) from the University of California—Santa Cruz. He undertakes analyses of non-state actors at the environmental science-policy interface. He conducts comparative analyses of media coverage of climate change, and analyses discourses that shape environmental science-policy interactions. He also examines the role of climate change-related celebrity endeavours, and explores links between these projects and environmental ethics, environmental social justice movements, and public understanding. Recent peer-reviewed publications include articles in *Transactions of the Institute of British Geographers*, *Climatic Change*, *Area*, *Geoforum* and *Global Environmental Change*.

Brian Doherty is a Senior Lecturer in Political Sociology at Keele University. His research focus is principally on environmental movements and he is the author of *Ideas and Actions in the Green Movement* (Routledge 2002) and co-editor (with Tim Doyle) of *Beyond Borders: Environmental Movements and Transnational Politics* (Routledge 2007).

Hugh Dyer is a Senior Lecturer in the School of Politics and International Studies (POLIS) at the University of Leeds. His first two degrees were taken at the University of Victoria and Dalhousie University in Canada, and his Ph.D. at the

London School of Economics. His research interests, teaching and publications are in the area of international theory and the theoretical implications of environmental change. He has published on the concept of environmental security and the significance of environmental values for international relations, as well as the normative aspects of biotechnology, and the role of normative theory in the study of international relations.

Andrew Jordan is Professor of Environmental Politics at the School of Environmental Sciences, University of East Anglia. He is particularly interested in the governance of environmental problems in different political contexts, but especially the European Union. He has edited *Environment and Planning C (Government and Policy)* since 1998, and authored or edited 10 books, including *Environmental Policy in Europe* (Ed., Routledge 2004), *The Europeanization of British Politics* (Ed., Palgrave 2006) and *The Coordination of the European Union* (Oxford University Press 2006).

Matthew Paterson is Professor of Political Science at the University of Ottawa. His most recent book is *Automobile Politics: Ecology and Cultural Political Economy* (Cambridge 2007), and his current research project is entitled 'climate change politics and the greening of the state'.

Hannes R. Stephan is a Ph.D. Candidate at the School of Politics, International Relations and the Environment (SPIRE), Keele University. His research interests include global and European environmental governance and he is currently working on the transatlantic conflict over agricultural biotechnology. His recent publications include a book chapter on transatlantic cultural politics and a co-authored article on European environmental leadership in the journal *International Environmental Agreements*. He is Technical Editor of the e-journal *In-Spire: journal of politics, international relations and the environment* (www.in-spire.org).

Tim Rayner works at the Tyndall Centre for Climate Change Research, University of East Anglia, as Senior Research Associate on the EU-funded ADAM project (Adaptation and Mitigation Strategies: Supporting European Climate Policy). Prior to this, he held various environmental policy research and teaching posts at the London School of Economics, University of Cambridge and University of Hull.

J. Timmons Roberts wrote his contribution to this volume while James Martin 21st Century Fellow at Oxford's Environmental Change Institute. His regular position is Professor of Sociology at the College of William and Mary, USA. His recent books include *A Climate of Injustice: Global Inequality, North-South Politics and Climate Policy* (with Bradley Parks, The MIT Press 2007), and *The Globalization and Development Reader* (with Amy Hite, Blackwell 2007).

Fariborz Zelli is Senior Research Associate at the Tyndall Centre for Climate Change Research, University of East Anglia. There, he is employed on the ADAM project (Adaptation and Mitigation Strategies: Supporting European Climate Policy). Since 2004 he has also been a Research Fellow of the Global Governance Project, for which he acts as co-ordinator of the MOSAIC research group ('Multiple Options, Solutions and Approaches: Institutional Interplay and Conflict'). Prior to joining the Tyndall Centre, he received a scholarship from the German Federal Foundation for Environment (DBU) and worked as a Research Assistant at the Center for International Relations/Peace and Conflict Studies at the University of Tübingen, Germany.

Abbreviations

a.m.	ante meridiem (before noon)
BAU	Business as Usual
CA	California
CBD	Convention on Biological Diversity
CD ROM	Compact Disc read-only memory
CDM	Clean Development Mechanism
CEO	Chief Executive Officer
CFC	Chlorofluorocarbon/s
CITES	Convention on International Trade in Endangered Species
CO	Colorado
CPR	Committee of Permanent Representatives
CSD	Commission on Sustainable Development
CT	Connecticut
CTE	Committee on Trade and Environment
DC	District of Columbia
DNA	Deoxyribonucleic acid
DSB	Dispute Settlement Body
EC	European Commission/European Community
ECOSOC	United Nations Economic and Social Council
EDA	Environmental Direct Action
EITs	Economy/Economies in transition
EJ	Environmental Justice
EMAS	Eco-management and Audit Scheme
EMG	Environmental Management Group
EMIT	Group on Environmental Measures and International Trade
EMS	Environmental Management Systems
Est.	Estimate/estimated
EU	European Union
FAO	Food and Agriculture Organization
FBI	Federal Bureau of Investigation
FCCC	Framework Convention on Climate Change
FEMA	Federal Environmental Management Agency
FoE	Friends of the Earth
FoEI	Friends of the Earth International
GATT	General Agreement on Tariffs and Trade
GDP	Gross Domestic Product
GEF	Global Environment Facility
GEG	Global environmental governance
GEMS	Global Environmental Monitoring System
GEO	Global Environmental Outlook

GHGs	Greenhouse Gases
GM	Genetically modified
GMEF	Global Ministerial Environmental Forum
ha	hectare
IAEA	International Atomic Energy Agency
IBRD	International Bank for Reconstruction and Development (World Bank)
IL	Illinois
IMF	International Monetary Fund
IMO	International Maritime organization
INFOTERRA	Global Environmental Information Exchange Network
IPCC	Intergovernmental Panel on Climate Change
IPE	International Political Economy
ISO	International Standards Organization
IUCN	International Union for the Conservation of Nature (World Conservation Union)
JI	Joint Implementation
JPI	Johannesburg Plan of Implementation
Kg	Kilogramme
KS	Kansas
MA	Massachusetts
MBIs	Market-Based Instruments
MD	Maryland
MDGs	Millennium Development Goals
MEAs	Multilateral Environmental Agreements
MLG	Multi-Level Governance
MNC	Multinational Companies
NATO	North Atlantic Treaty Organization
NEPIs	New Environmental Policy Instruments
NGO	Non-governmental Organization
NJ	New Jersey
ODA	Official Development Aids
OECD	Organisation for Economic Co-operation and Development
PA	Pennsylvania
PRSP	Poverty Reduction Strategy Papers
SUV	Sports Utility Vehicle
SWMTEP	System-Wide Medium-Term Environmental Programme
TEU	Treaty on European Union
TRIPS	Trade-related aspects of intellectual property rights
TX	Texas
UN	United Nations
UNCED	United Nations Conference on Environment and Development
UNCHE	United Nations Conference on the Human Environment
UNCLOS	United Nations Conference on the Law of the Sea
UNDP	United Nations Development Programme

UNEP	United Nations Environmental Programme
UNESCO	United Nations Educational, Scientific and Cultural Organization
UNFCCC	United Nations Framework Convention on Climate Change
UNFPA	United Nations Population Fund
UNGA	United Nations General Assembly
UNIDO	United Nations Industrial Development
US(A)	United States (of America)
VAs	Voluntary Agreements
vs.	versus
WCED	World Commission on Environment and Development
WHO	World Health Organization
WMO	World Meteorological Organization
WRI	World Resources Institute
WSSD	World Summit on Sustainable Development
WTO	World Trade Organization
WWF	World Wide Fund for Nature

Essays

Essays

Globalization: The Environment and Development Debate

J. Timmons Roberts

INTRODUCTION

International debate on and politics of the environment can be seen as the result of a tension between development goals and environmental protection. Over the last few decades, humanity has seen huge increases in economic development (e.g. Norberg 2007). This has brought huge increases in global GDP, and overall great increases in indicators of health, education, and physical well-being. The most recent phases of international development have also taken place while international trade has swung rapidly upward, part of what observers have called the unstoppable trend of 'globalization' (e.g. Roberts and Hite 2007). However, this pace of development and globalization has led to immense environmental degradation, and for over three decades many scientists have been arguing that the scale of this degradation threatens humanity. Yet most of the debate on development has for years ignored the environmental impacts that potentially undermine the entire model being held up as an ideal for the poor nations to emulate: endless economic growth and industrialization (Sutcliffe 2000).

Along with economic globalization has come the globalization of environmental damage. The crux of the problem is that while economic development and globalization threaten the environment, many scholars and politicians argue that development is the key to achieving decent human well-being and to solving these same environmental problems (e.g. World Bank 1992; Grossman and Kreuger 1995; Selden and Song 1995). As Mexican President Carlos Salinas was reported to have said on US television, Mexico needed first to pollute itself in order to meet its basic needs, and then environmental concerns could be addressed later on. The core issue then is how to reconcile the two goals of economy and ecology. Much research and policy is based on the premise that they can be brought together, but this may be wishful thinking.

The third 'leg' of the tripod of sustainability, of course, is equity: how to harmonize development and environmental sustainability with unequal development and international injustice. Thirty years of negotiations, high-level reports, and 'Grand Bargains' leave us still grasping for solutions that will bring these three elements into alignment. With an admirable 'bias for hope', much previous work assumed that by dealing with equity and justice we would solve problems of environmental sustainability (Dobson 2003; Agyeman et al. 2003). There may be a level of denial involved, however, since there is no reason to assume that democratic or just solutions will solve environmental crises. Assuming so is nearly as dangerous as assuming authoritarian or highly unequal societies will have the means and desire to solve environmental problems, or that economic growth will automatically do so.

In this essay, I describe some of the debate between environment and development, going back to the first UN Conference on Environment and Development, held in Stockholm in 1972. This story lays some of the groundwork for this book, and shows how negotiations on environment will always founder and fail if global inequality and the development needs of poor nations continue to be treated as secondary: again and again, we learn that they cannot be treated in isolation. I then spend some time discussing globalization and what it means, and tying globalization to environmental damage. I describe the difference between globalization and development, and where this leaves development planners and the politics of development today. The conclusion includes some of my thoughts on what will be needed for popular control of the social and environmental impacts of globalization. There are no easy solutions here—the tension between development and environmental protection is profound. It is, however, a tension we cannot afford not to solve.

THE ENVIRONMENT AND DEVELOPMENT DEBATE: WHOSE PRIORITIES GET ADDRESSED? CAN GROWTH AND ENVIRONMENT BE ALIGNED IN 'SUSTAINABLE DEVELOPMENT'?

Fears of ecological crises and doomsday scenarios about the shrinking resource base, spreading pollution, and ever-expanding populations (Carson 1962; Meadows et al. 1972; Ehrlich 1970) drove the rapid growth of a vital citizens' environmenatl movement in the 'global North'. The Swedish Government responded by submitting a proposal to the UN in 1970 declaring an 'urgent need for intensified action at the national and international level, to limit and, where possible, to eliminate the impairment of the human environment'. The UN Economic and Social Council (ECOSOC) approved the idea and sent it to the General Assembly, which endorsed it and convened 'the UN Conference on the Human Environment'. In 1972 113 nations met in Stockholm, Sweden, where conference chair Maurice Strong told thousands of protesters outside the event that they were successfully pushing the nations of the world to address the issue.

To the surprise of most environmentalists and delegates from nations from the global North, Brazil and India led a vigorous resistance to their agenda. India argued that there are different pollutions of the poor and rich, 'the wealthy worry about car fumes; we worry about starvation'. The conference put the North on a head-on collision course with priorities of nations in the global South, which saw poverty driving environmental problems: soil erosion, deforestation, desertification, and diminishing water resources. The core idea was that basic human needs must come before the ultimately distracting work of environmental clean-up. The Sri Lankan ambassador said plainly: 'we must not, generally speaking, allow our concern for the environment to develop into a hysteria'.

Brazil argued that rich nations could not tell it to stop developing, saying essentially if you want us to clean up, you must help us to develop first, so we can clean up later. 'No growth' or even 'limits to growth' were absolutely unacceptable to these countries, given their terrible situations and position far behind the wealthier nations. In fact, environmental problems in poor countries were argued to be due not to overuse, but insufficient use of resources (Guimarães 2000). Developing nations took huge offence at Northern concern about their population growth rates, since it implied outside meddling in deep cultural values, unfair judgments about 'crowding' and 'overpopulation', and a selfish fear of migrants moving north (Galeano 1973). Further, developing nations made clear from the start that they had no money to spare to address the issue: if they wanted the issues dealt with, the richer countries should pay for the clean-up. It was a 'happy coincidence', Brazilian delegates argued, that rich countries have the biggest environmental problems, cared the most about them, and had the fiscal resources to clean them up (Guimarães 2000). Of great concern was that wealthy nations might use concepts of 'our Spaceship Earth' and international environmental treaties to limit their sovereign rights to their own resources, which they argued were not global, but national. Southern nations argued that those in the North should stop claiming that all nations have a share of the Earth's resources which form a 'common pool' or 'World Trust': this 'beautiful assumption' requires sharing also of economic and political power, industry, and financial control, which wealthy nations found quite unthinkable (Guimarães 2000).

Many other developing nations endorsed Brazil's positions at Stockholm, and reaching any consensus there was largely the result of reducing demands and adding the promise of financial aid to the South to help with addressing environmental issues. At the June preparatory conference to Stockholm in Founex, Switzerland, a compromise was made: Northern nations admitted that their pollution was impinging on poor nations' development, it was accepted that poverty was the overriding cause of environmental destruction in poor countries and that economic growth will be needed there, and national sovereignty was guaranteed as a fundamental principle of any agreement.

What emerged was a formal statement, the Stockholm Declaration, which contained language for both those nations wanting an acknowledgement of the gravity of environmental issues, and those wanting assurance that this acknowledgement would not hamper their growth. The declaration included 26 principles on everything from environmental education and law to urbanization and population. Principle 1 began with 'the fundamental right to freedom, equality and adequate conditions of life', a series of issues which imply economic development but have remained contentious up to the present. Principle 9 stated explicitly that, 'Environmental deficiencies generated by the conditions of underdevelopment and natural disasters pose grave problems and can best be remedied by accelerated development.' It continued that this development would come 'through the transfer of substantial quantities of financial and technological assistance' in addition to national efforts in poorer nations. Principle 23 explicitly warned against instituting 'standards which are valid for the most advanced countries but which may be inappropriate and of unwarranted social cost for the developing countries'.

The 'Action Plan for the Human Environment' agreed upon at Stockholm had 109 recommendations: for the scientific assessment of environmental problems, for ambitious institution-building for their management, and for a new UN Environment Programme (UNEP) to co-ordinate all UN actions on the environment. However, the UNEP was based in Nairobi, Kenya, which for a series of reasons caused the agency to struggle and never to function as a co-ordinating agency (Reed 1992; Ivanova 2006). The principles were not the basis of international laws, and the 109 recommendations were largely overlooked. However, Reed (1992) argues that Stockholm 'successfully crystallizes the underlying issues' of global environmental politics, and forged a fragile compromise for future actions on the environment, such as a series of international treaties which were developed over the next decades.

The 10-year follow-up to Stockholm was the 1982 UN conference in Nairobi, which hardly anyone remembers any more. It was described as 'a complete disaster', as US President Ronald Reagan's delegate 'succeeded in completely sabotaging the conference' (Vaillancourt 2000). To pick up the pieces, in 1983 the UN General Assembly set up a World Commission on Environment and Development (WCED) which became known as the Brundtland Commission (under Ms Gro Harlem Brundtland of Norway). It included six Westerners, three East Europeans, and 12 representatives of developing nations. In 1987, the commission issued the influential report 'Our Common Future', which popularized the term 'sustainable development'. It defined the term as 'Development that meets the needs of the present without compromising the ability of future generations to meet their own needs' (p. 43). The report's authors conceptualized sustainable development as having three dimensions: economic-developmental, environmental-ecological, and social-political (equity) (Vaillancourt 2000).[1]

The concept of 'sustainable development' gained immediate and wide acceptance, partly because it was so vague and sounded so good that everyone could agree with it (Clapp and Dauvergne 2004; Humphries, Lewis and Buttel 2002). The Brundtland report allowed all sides to embrace this concept of sustainable development, including the business community. 'The chemical industry can help to make sustainable development a reality,' said Dow Chemical's CEO Frank Popoff. The American Chemistry Council's Vice-President Terry Yosie said that, 'Sustainable development is an appropriate framework for integrating the environmental, economic, and social issues that are facing industry and society . . . We endorse that framework' (Chemical Week, 2002). On the need to get business on board, Maurice Strong said in 1992: 'The environment is not going to be saved by environmentalists. Environmentalists do not hold the levers of economic power' (Bruno and Karliner 2002: 22).

However, the concept of sustainable development was problematic from the start. It seemed to mean whatever people wanted it to mean: everyone had something different in mind when claiming it as their practice or visualizing the future with it as their goal. People are still debating about what it means: David Pearce a decade ago collected 13 pages of definitions of sustainable development, all sharing and emphasizing differently the three parts of economic growth, environmental protection, and equity. Sociologists Craig Humphrey, Tammy Lewis, and Fred Buttel wrote that sustainable development, in the Brundtland report and after, ignored any 'limits to growth', assuming that environmental protection and economic growth were at least potentially entirely complementary. It obtained its rapid adoption and seeming instant consensus by 'instead focusing on how sustainable development can be achieved' (Humphrey, Lewis and Buttel 2002, p. 222). They concluded that the sustainable development approach is thus inherently 'pro-technology, pro-growth, and compromise oriented' (Humphrey, Lewis and Buttel 2002, p. 222). It fits well, they argued, with the technocratic approach to environmental crises, that these crises can be resolved by scientific management.

'THE GRAND BARGAIN AT RIO'

After years of negotiating and four crucial preparatory conferences, the Rio Earth Summit (UN Conference on Environment and Development) in 1992 was a watershed event, with 168 countries represented, 117 heads of state, and a huge parallel 'People's Forum'. The huge numbers of non-governmental organizations (NGOs) at the event took a two-pronged approach, working inside the convention hall as they could through their country delegations, and outside they built networks and collaborations around the world. In this way, the event effectively globalized the environmental movement (of course with distinct national, regional, and international inequalities) in which new regions were represented and given voice in the new and more global networks.

Sensing key risks and the potential for widespread anti-industry attitude and a potential new wave of global regulations, industry had a huge delegation at Rio in 1992 and at the next big summit in Johannesburg in 2002, and sought actively to redirect the debate. In fact, industry funded many of the conferences, and was well placed in major discussions (Vaillancourt 2000; Humphrey, Lewis and Buttel 2002; Bruno and Karliner 2002). This explains also the emergence of voluntary approaches emphasizing the sustainable development solution.

Coming from Rio were several key treaties and documents, including the Convention on Biological Diversity, the Framework Convention on Climate Change, the Document on Forest Principles, and the Rio Declaration. A massive consensus document called Agenda 21 attempted to lay out how to attain a strong economy and environmental protection for all people on Earth. An 800-page plan of action, with more than 100 recommendations, Agenda 21 addressed environmental protection, socio-economic development, equity, justice and nearly everything else. Vaillancourt (2000) describes its 'drab style', as it was written by many technocrats from around the world, edited and re-edited, and condensed into summaries. These suggestions were not legally binding, and except for the 'Local Agenda 21' approach for municipal planning, most have been ignored and forgotten.

All five documents from Rio reflected the deep North–South split over environment and development. To resolve it, a 'Grand Bargain at Rio' emerged in which poor countries agreed to work on environmental protection if their growth was secured and if any costs of their cleaning up were borne by the wealthy. Promises of huge amounts of aid to address global environmental issues like ozone and climate change were made (Hicks et al. 2008). Chapter 33 of Agenda 21 reads, '[t]he implementation of the huge sustainable development programs will require the provision to developing countries of substantial new and additional financial resources'. The price tag placed on achieving sustainable development in the Third World was estimated at US $561,500m. a year, with the global North expected to bankroll $141,900m. in low or no-interest 'concessional' assistance (or 20% of the total cost). Developing countries were expected to foot the rest of the bill. Of the $125,000m. in concessional assistance, about $15,000m. a year was supposed to be devoted to global environmental issues, with the rest targeted at sustainable development programmes in developing countries (Robinson 1992).

Making clear that environmental issues were only secondary to economic development and other needs, the group of developing nations declared that 'a special Green Fund should be established to provide adequate and additional financial assistance to them'. The key point was that funds for environmental protection should not be taken from existing development aid. Specifically, they proposed a special 'Earth Increment' to be added to the World Bank's development assistance funds, to 'provide virtually free environmental aid to the very poorest nations'. The World Bank was designated as the home for the Earth Increment, as a major fund for environmental protection called the Global Environment Facility (GEF) was founded. The Earth Increment was to be a 15% boost in Official Development Assistance (ODA) funding, to help with implementation of the Agenda 21 plans.

The GEF Charter said that the agency was to only deal with 'global' environmental issues, not ones of local concern. According to its founding charter, projects funded by the GEF also had to be 'additional', that is, not likely to have taken place without outside funding. Years of battles have ensued over which projects count as addressing 'global' versus local environmental issues, and 'additionality' has proven nearly impossible to prove. These struggles continue, and many developing nations remain resentful of having the World Bank as the administrator of the GEF and other environmental funds (Müller 2007; Okereke and Mann 2007; Hicks et al. 2007).

The UN Framework Convention on Climate Change (UNFCCC) was a key outcome of Rio, and led in 1997 to the Kyoto Protocol. There are three points in the UNFCCC that reveal the enduring tensions between North and South over the issues of development and environment. The objective of the treaty, outlined in Article 2, is to achieve 'stabilization of greenhouse gas concentrations in the atmosphere at a level that would prevent dangerous anthropogenic interference with the climate system. Such a level should be achieved within a time-frame sufficient to allow ecosystems to adapt naturally to climate change, to ensure that food production is not threatened and to enable economic development to proceed in a sustainable manner.'[2] The convention places the heaviest burden for fighting climate change on industrialized nations in Article 3.1. 'The Parties should protect the climate system for the benefit of present and future generations of humankind, on the basis of equity and in accordance with their *common but differentiated responsibilities and respective capabilities*. Accordingly, the developed country Parties should take the lead in combating climate change and the adverse effects thereof' (emphasis added). And importantly, it describes how industrialized nations agree under the convention to support climate-change activities in developing countries by providing financial support with grants and loans through the convention and managed by the GEF above and beyond any financial assistance they already provide to these countries (UN 2007).

Built into the treaties was a profound inequality in interests, in part because of the unequal representation of scientists and negotiators at Rio and the other earth summits. At the sixth Conference of the Parties of the UNFCCC, for example, the USA brought 99 formal delegates and the European Commission brought 76, while many small island and African states were lucky if they could put together a one-, two- or three-person delegation (Richards 2001). The problem is reflected as well in the number of authors drafting policy and key documents. In the 1995 Intergovernmental Panel on Climate Change Working Groups I, II, and III, out of 512 Working Group I authors, 212 were US citizens and 61 were from the United Kingdom, while only 12 authors came from India and China combined. Working Groups II and III also showed very wide inequalities in representation (Kandlikar and Sagar 1999). There is little question that these documents would look very different if written with equal representation from the global South. One of the clearest differences would likely be that the right to development would gain equal attention with environmental protection.

This leads an inquisitive observer to ask the most basic question: why did poor countries accept Kyoto? Atiq Rahman, executive director of the Bangladesh Center for Advanced Studies, reported quite starkly: 'The Kyoto Protocol had almost nothing to do with developing countries. It was a negotiation between the OECD countries on their agreed allocations, on how to reduce their greenhouse gases. On the last night, the developing countries were brought in to talk about it and to accept it. ... I have done an interview of 85 leaders of the negotiations just after the Kyoto Protocol and asked them why they signed this stupid document, which is totally iniquitous. And they said, 'because they will set up an adaptation fund, [and] there could be some money in it'. He continued, 'As I said, we are in two different worlds. Now [developing country negotiators] are saying, "the Kyoto Protocol, it's so unjust, totally iniquitous, but for the sake of negotiation, let us accept it"'' (Rahman 2001).

The forceful reaction of the USA to such a shift in emphasis makes clear the tension at incorporating 'the right to development' in environmental agreements. Before the Rio Declaration was adopted at the 19th Plenary on 14 June 1992, the US delegation submitted an 'interpretative statement expressing reservations'

9

(Sand 1994) on this and a few other articles. The point was made again at the 2002 World Summit on Sustainable Development in Johannesburg. It read that, 'The U.S. does not, by joining the consensus on the Rio Declaration, change its longstanding opposition to the so-called "right to development". Development is not a right. On the contrary, development is a goal we all hold. The U.S. cannot agree to, and would disassociate itself from, any interpretation of Principle 3 that accepts "a right to development" or otherwise goes beyond that understanding ... The United States does not accept any interpretation ... that would imply a recognition or acceptance by the United States of any international obligations or liabilities, or any diminution in the responsibilities of developing countries' (UNCED 1992a; USA 2002).

This exchange, decades after the very similar debates we saw at the Stockholm summit, shows how issues of development persistently 'spill over' onto environmental ones (Roberts and Parks 2007). Rich nations have made heroic efforts to compartmentalize climate issues and development issues, but poor nations' development concerns are not going away. A dozen years after Rio we are still fighting the same battle, which by now we should understand: that environment and development are inseparable and we must better understand the world views and beliefs of the world's poorer nations if they are to participate in meaningful efforts to address climate change. As designing a workable climate regime for after Kyoto expires in 2012 weighs ever heavier on the minds of Western policy-makers, the need to integrate these two policy objectives has become unavoidable.

GLOBALIZATION AND THE COMPLEXITY OF THE NEW INTERNATIONAL DIVISION OF LABOUR FOR THE ENVIRONMENT/DEVELOPMENT DEBATE

An iconic image of globalization is a massive container ship, bringing cheap toys, electronics, clothing and furniture halfway around the world. This choice of icons is well supported: about nine-tenths of world trade of goods is by ship, and most of the value is now carried in containers (People's Daily Online 2004). Staring at a shipload of containers is also like gazing at the Sphinx: its meaning is entirely obscured from our view. Hidden in each corrugated steel box are thousands of products produced around the world with 'stuff' mined, farmed or pumped from far corners of the Earth. The wages, working conditions, and lives of workers whose labour is embodied in those products are organized differently in varied types of workplaces and communities, under the purview of either local, national, or international firms, under formal work rules and contracts or in entirely informal arrangements, in small family-owned, publicly-held corporations, or state-owned enterprises. The same products could be produced under strict or extremely lax environmental and health conditions. All these variables can differ even within nations: for example, in the first half of 2006, foreign-owned export-processing firms generated US $465,300m. of China's total foreign trade (58%), while state-owned enterprises earned $195,300m. and Chinese private firms earned \$135,100m. in trade (Qi 2006). Each firm may differ in its strategy of production and environmenatl practice, but facing global competition, we know they all share one criterion: they are required to keep production costs down.

Manufacturing for export has been a prime development strategy of nations attempting to move up in the global hierarchy of nations and gain the fruits of modernization. Industrialization has for decades been seen as synonymous with 'development'. So right now, it seems that China's dominant manufacturing sector is powering its success in a global competition for development.

Many environmentally-concerned consumers are now asking retailers for labels on products comparing their complete 'carbon footprint' or 'ecological footprint' to other similar products. However, the environmental implications of global production and global sourcing are massive, and extremely complex. To understand them, we need to understand the 'commodity chain' running from the place where the raw material was extracted, to its transport, processing, marketing, consumption, and final disposal (Gereffi and Korzeniewitz 1995, etc.). With cheaper transportation systems, the production, consumption and disposal of goods can be done in many locations and organized very differently. Even services have commodity chains and ecological footprints that often reach around the globe, as they rely on workers, equipment and energy produced in diverse locales. Environmental impacts and risks to workers and communities are great at certain nodes in these chains, and the social development benefits are often concentrated at very different nodes.

Building on dependency and world-systems theories (which focused on the labour basis of value), sociologist Stephen Bunker pioneered the idea that extraction of resources degrades both natural resources and the societies in those regions, while enriching the 'core' and especially urban parts of the world economic system (Bunker 1985; see also Galeano 1993 and other dependency theorists). This was an old but under-studied type of 'unequal exchange' between the rich and poor regions of the world: each cycle makes the developing nations more desperate to extract greater volumes of what they can sell on global markets.

The debt crisis of the 1980s (which faced many nations that borrowed heavily in the petrodollar boom days of the 1970s) is frequently considered a prime way that globalization has caused deeper environmental damage in the extractive regions of the world (e.g. Reed 1992). Martin Khor described a broad set of resources flowing from South to North (1994). Joan Martínez-Alier and his colleagues have now advanced the idea of 'environmentally unequal exchange' and an 'ecological debt' which is owed by the North as a result of its unfairly profiting at the expense of the South (2001, 2002, etc.). These ideas have finally begun to be quantitatively analysed and confirmed by comparisons of impacts and benefits in different world regions (Giljum and Eisenmenger 2004; Eisenmenger, Martin and Schandl 2007; Fischer-Kowalski and Amman 2001).

The poorest nations of Africa and Latin America extract natural resources and 'mine the soil' for greater export cash crop harvests. However manufacturers like China and India (and many others) are increasingly doing much more than that: they are assembling components or doing final assembly of manufactures which were until very recently done in the wealthy nations of the global North. The 'new international division of labour', first described by Fröbel, Heinrichs and Kreye in 1981, has now seen a nearly complete rearrangement of the factories to nations with low labour costs. The role of other factors like low environmental regulations and cheap energy has been much debated.[3]

However Arrighi et al. (2003) decisively document that simply adopting industrialization as a development model in no way guarantees improvement in living standards. Rather, with the dropping of 'innovation rents' in later stages of the product cycle, poor nations gain decreasing long-term benefit from assembly plants (Dicken 2007). Rather, having cheap labour means that few benefits will accrue locally. As Gary Gereffi and others have documented, the control of the 'commodity chains' by name-brand firms guarantee them far larger proportions of the profits, of which quite little goes to the actual manufacturers, who are often sub-sub contractors (Gereffi and Korzeniewicz 1994).

The main point I wish to make from this discussion is that development efforts on either an extractive or manufacturing model bring environmental damage, and often without securing social improvement. That is, countries rushing to these models risk gaining neither environmental protection nor development. Both could potentially be done with reasonable protection of the human and natural environment, but combining them into some 'sustainable development' is not easy nor particularly likely in the current global climate. Some other model is required.

GLOBALIZING CONSUMPTION, AND WHAT MIGHT BE NEEDED TO MOVE TOWARDS A POSITIVE GLOBALIZATION

Is globalization the prime cause of the destruction of the global environment? Or is it potentially the way towards addressing what are now global environmental problems? Globalization is more than the flow of goods, of course: there is services trade, cultural globalization, financial globalization, biological globalization, migration, and political globalization. Each has the potential to act as a brake on or an enhancer of economic globalization and its environmental impacts. For example, the spread of environmentalist ideas could drive a new wave of regulations at the national and international level to redirect the global economy. But as a way to wrap up this essay, I would like to consider an example.

A concrete example of globalization and its environmental and social impacts would be to trace the rise of fast food. Global brands like McDonald's or Kentucky Fried Chicken are building global franchises and displacing locally-produced and traded products every day. The main products these establishments are selling offer six main characteristics: low price, consistency, the biological attraction of fat, sugar and salt, and the glamour of modernity. One company, Yum! Brands, Inc., based in Louisville, Kentucky, USA, 'is the world's largest restaurant company in terms of system restaurants, with over 34,000 restaurants in over 100 countries', in 2005 generating over US $9,000m. in total revenue (Yum! 2007). In the third quarter of 2006, Yum! China Division alone had more than 2,400 system restaurants (Q3 2006), and the firm was opening three restaurants a day around the world.

To open these restaurants, franchise owners and managers need to pay substantial monthly rent and fees to the corporation, and follow extremely detailed methods of management, production, sourcing, publicity, and almost everything else (Schlosser 2001). The homogenization of textures and flavours of food requires that crops be uniform, and heavily processed in closely-supervised factories. Fast food animals (like many others in 'modern production') are often fattened on soybeans or other feed grains, which to meet demand are sometimes grown in areas previously in rainforests, savanna, or used to grow food for local human consumption. The social and environmental profile of agriculture is being transformed, as large-scale production of animals and crops is heavily favoured in sourcing by fast food chains over local family farming (e.g. Magdoff et al. 2000). High levels of chemical inputs are needed to maintain soil fertility, control pests, and to protect animal health. Large amounts of energy and inputs are needed for transport, processing and packaging. Less discussed is the loss of local foodways, in which people have built their local cultural traditions around food production and consumption over centuries or millennia, and which tie people to their place and its protection. Finally, there is the obesity epidemic which is spreading from the USA along with these high-fat, high-sugar foods.

So from this one simple example one can see many elements of the arguments raised by activists and observers lumped into the 'anti-globalization' camp. These are not all blamed on one brand or even solely on fast foods, but on the whole complex of changes in food systems going on in developing countries today. First, that globalization changes the way people make decisions, so it reorients local labour, land and other resources towards meeting the needs of the market, especially the global market, at the expense of local subsistence and improvement. Second, this loss of local markets for smallholder-grown products can drive migration from rural to urban residence, which creates another cascading set of social and environmental impacts in both sending and receiving locations. Third, global production and distribution means massive increases in the energy needed to transport products around, when compared to earlier systems of local provisioning. Fourth, globalization is increasing overall levels of consumption, which, when combined with growing world populations, means increasing damage to ecosystems. Fifth, globalization increases risks by decreasing diversity in native food systems, with the adoption of homogenous hybrid and genetically-modified crops. Sixth, the current wave of globalization appears to be increasing inequality both within nations and between them, as local entrepreneurs or state politicians profit from working with transnational corporations or investors, while others gain only seasonal and menial work paying barely survival wages, and may lose other resources such as access to communal lands. And seventh, under pressure to increase their exports to gain hard currency to pay off international debts, developing nation governments look for resources to export, and often these have devastating impacts on rainforests and native peoples in these ecosystems.

While free market advocates rightly point out globalization's increasing global wealth overall, and the security of food supply that can come with trade and aid, these arguments animate a growing and sometimes converging movement of activists fighting for social justice, workers' rights, fair trade, and the environment. Just to mention a few, the Chipko movement in India is fighting deforestation there by planting millions of trees. Local environmental and peasant farmers groups are springing up all over China. Environmentalists and labour unions have joined together occasionally to fight free trade treaties and the World Trade Organization. Indigenous people in Colombia, Ecuador and Nigeria have fought big oil companies like Occidental Petroleum and Shell, attempting to force them to clean up their environmental impacts on their local lands. As mentioned, the new wave of environmentalism seeks to address environmental crises through labelling of products, sometimes with their country of origin, with the 'food miles' they have travelled, or their 'carbon' or 'ecological footprint'. People affected by dams are becoming more organized and linked internationally. The list is growing, and some authors argue that there are increasingly strong 'transnational advocacy networks' which allow local peoples more effectively to fight national or international actors like corporations, governments, or international financial institutions like the IMF and the World Bank (e.g. Keck and Sikkink 1997).

One can find support in these trends for a 'bias for hope', and we cannot afford to not be hopeful. However, one can also quickly identify profound weaknesses of the NGO-led model for addressing the environmental impacts of development under the current model of globalization. Social-environmental NGOs are notoriously ephemeral, growing rapidly and then dispersing, or being deradicalized or co-opted by governments or professionalization of their leaders. 'Compassion fatigue' makes it ever more difficult to mobilize protests in the wealthy nations, and even to secure long-term commitments to foreign assistance that might address environmental and social problems caused by development. The Grand Bargain at Rio is threatening to collapse. Global treaties to protect environmental-social systems (such as the Convention on Biological Diversity or the Kyoto Protocol) are notoriously excruciating to negotiate through the international system, and almost entirely lack effective enforcement mechanisms.

So it is clear that stronger and more effective global institutions are required to balance the need for economic development with the equally important need to protect the environment that sustains development. For nation-based politicians to support (giving power over to) such global institutions requires their being able to endure pressure at the national and local levels, which depends on their gaining local support for these efforts. Local support in turn requires strong social movements, which in turn require an open, democratic political atmosphere (as differently defined around the world). The basic conditions for participatory democracy, in turn, are the rule of law, transparency in government, constitutional sharing of power, freedom of association, the press and expression, and so on. So I would argue that to achieve these underlying conditions for combining the three elements of sustainable development—environment, equity and economy—requires revisiting the core issues of the bargain at Rio and before that, the core conflict at Stockholm. Development pathways, ecologically-unequal exchange, and the ecological debt need all to be addressed as the next international division of labour continues to unfold in new ways. The picture is complex but clear: that to address environmental issues requires we understand development.

Bibliography

Agyeman, Julian, Robert Bullard and Bob Evans (Eds). 2003. *Just Sustainabilities: Development in an Unequal World*. Cambridge, MA: The MIT Press.

Arrighi, G., B. Silver and B. Brewer. 2003. Industrial convergence, globalization and the persistence of the North-South divide. *Studies in Comparative International Development* 38 (1): 3–31.

Boyd, Emily, Nathan E. Hultman, Timmons Roberts, Esteve Corbera and Johannes Ebeling. 2007. The Clean Development Mechanism: Current Status, Perspectives and Future Policy. Environmental Change Institute/Tyndall Centre for Climate Research. Working Paper, June 2007.

Bruno, K., and J. Karliner. 2002. *Earthsummit.Biz: The Corporate Takeover of Sustainable Development*. Corpwatch and Food First Books: San Francisco and Oakland.

Bunker, Stephen. 1985. *Underdeveloping the Amazon: Extraction, Unequal Exchange and the Failure of the Modern State*. Urbana, IL: University of Illinois Press.

Carson, Rachel. 1962. *Silent Spring*. Boston: Houghton-Mifflin.

Clapp, Jennifer, and Peter Dauvergne. 2004. *Paths to a Green World: The Political Economy of the Global Environment*. Cambridge, MA: The MIT Press.

Dicken, Peter. 2007. *Global Shift: Mapping the Changing Contours of the World Economy*. Fifth Edition. London: Sage Publications; New York: Guilford Press.

Dobson, Andrew. 2003. "Social Justice and Environmental Sustainability: Ne'er the Twain Shall Meet?", in Agyeman, Julian, Robert Bullard and Bob Evans (Eds). *Just Sustainabilities: Development in an Unequal World*. Cambridge, MA: The MIT Press.

Ellis, J., H. Winkler, J. Corfee-Morlot, and F. Gagnon-Lebrun. 2007. CDM: Taking Stock and Looking Forward. *Energy Policy* 35: 15–28.

Fischer-Kowalski, Marina, and Christof Amman. 2001. Beyond IPAT and Kuznets Curves: Globalization as a Vital Factor in Analyzing the Environmental Impact of Socio-Economic Metabolism. *Population and Environment* 23 (1): 7–47.

Fröbel, Folker, Jürgen Heinrichs, and Otto Kreye. 1981. *The New International Division of Labor*. New York: Cambridge University Press.

Humphrey, Craig R., Tammy L. Lewis and Frederick H. Buttel. 2002. *Environment, Energy, and Society: A New Synthesis*. Belmont, CA: Wadsworth.

Meadows, Donella, Dennis Meadows, Jorgen Randers and William Behrens. 1972. *Limits to Growth*. New York: Universe Books.

Ehrlich, Paul. 1970. *The Population Bomb*. Boston: Ballantine Books.

Eisenmenger N., J. Ramos-Martin and H. Schandl. 2007. Transition in a changed context: patterns of development in a globalizing world. In M. Fischer-Kowalski and H. Haberl (Eds), *Socio-ecological Transitions and Global Change: Developments in Societal Metabolism and Land Use*. Cheltenham, Edward Elgar. pp. 179–222.

Galeano, Eduardo H. 1971 (English 1973). *Open Veins of Latin America: Five Centuries of the Pillage of a Continent*. New York: Monthly Review Press.

Gereffi, Gary, and Miguel Korzeniewicz (Eds). 1994. *Commodity Chains and Global Capitalism*. Westport, CT: Praeger.

Giljum, S., and N. Eisenmenger. 2004. North-South trade and the distribution of environmental goods and burdens. A biophysical perspective. *Journal of Environment and Development* 13(1): 73–100.

Grossman, Gene M., and Alan B. Krueger. 1993. "Environmental Impacts of a North American Free Trade Agreement", in P. Garber (Ed.), *The U.S.-Mexico Free Trade Agreement*. Cambridge, MA: The MIT Press.

Guimarães, Roberto. 2000. Brazil and Global Environmental Politics: Same Wine in New Bottles?, *Sustainability and Unsustainability on the Road from Rio*, presented at Research Committee 24 of the International Sociological Association conference, Rio de Janeiro, August 2000.

Hicks, Robert, Bradley C. Parks, J. Timmons Roberts, and Michael J. Tierney. 2008. *Greening Aid? Understanding Foreign Assistance for the Environment*. Oxford: Oxford University Press.

Ivanova, Maria. 2006. *Understanding UNEP: Myths and Realities in Global Environmental Governance*. Ph.D. dissertation: Yale University.

Kandlikar, Milind, and Ambuj Sagar. 1999. Climate Change Research and Analysis in India: An Integrated Assessment of a South-North Divide. *Global Environmental Change* 9 (2): 119–138.

Keck, Margaret E., and Kathryn Sikkink. 1997. *Activists beyond borders: Advocacy Networks in International Politics*. Ithaca: Cornell University Press.

Khor, Martin. 1994. "South-North resource flows and their implication for sustainable development." *Third World Resurgence*, No. 46, pp. 4–25, 1994.

Magdoff, Fred, John Bellamy Foster, and Frederick Buttel (Eds). 2000. *Hungry for Profit: the Agribusiness Threat to Farmers, Food, and the Environment*. New York: Monthly Review Press.

Muradian, R., M. O'Connor and J. Martínez-Alier. 2002. Embodied Pollution in Trade: Estimating the 'Environmental Load Displacement' of Industrialised Countries. *Ecological Economics* 41(1): 41–57.

Martínez-Alier, J. 2003 *The Environmentalism of the Poor: A Study of Ecological Conflicts and Valuation*. Cheltenham: Edward Elgar.

Müller, Benito. 2007. Nairobi 2006: Trust and the Future of Adaptation Funding. Oxford Institute for Energy Studies, Working Paper EV38.

Nobre, Carlos. 2007. Comments in Public Forum, Climate Change and the Future of the Amazon Conference, University of Oxford, Environmental Change Institute, March 20–22, 2007.

Norberg, Johan. 2007. "In Defense of Global Capitalism," in J. Timmons Roberts and Amy Hite, *The Globalization and Development Reader*. London: Blackwell.

People's Daily Online. 2004. "Huge room for China to develop containers, says shipping giant." Online at http://english.people.com.cn/200408/13/eng20040813_152880.html

Qi, Wu. 2006. "China's Surging Foreign Trade: Joy Tempered with Sorrow (29/09/2006)." Web Features, Embassy of the People's Republic of China in the United Kingdom of Great Britain and Northern Ireland. Online at http://www.chinese-embassy.org.uk/eng/zt/Features/t274358.htm.

Reed, David (Ed.). 1992. *Structural Adjustment and the Environment*. Boulder: Westview Press.

Roberts, J. Timmons, and Nikki D. Thanos. 2003. *Trouble in Paradise: Globalization and Environmental Crises in Latin America*. New York: Routledge.

Robinson, N. A. (Ed.). 1992. *Agenda 21 and UNCED Proceedings*, Volumes 1 and 2. New York: Oceana Publications.

Sand, Peter. 1994. International Environmental Law After Rio. *European Journal of International Law*, 377–389.

Schlosser, Eric. 2001. *Fast Food Nation: The Dark Side of the All-American Meal*. Boston: Houghton Mifflin.

Selden, T. M., and A. Song. 1995. "Neoclassical Growth, the J Curve for Abatement and the Inverted U Curve for Pollution." *Journal of Environmental Economics and Management* Vol. 29: 167–168.

Sissel, Kara. 2002. "Brazil and Chile Begin Third-Party Verification." *Chemical Week* July 3/10: 63.

Sutcliffe, Bob. 2000. "Development After Ecology," in *From Modernization to Globalization: Perspectives on Development and Social Change*, edited by J. Timmons Roberts and Amy Hite. Malden, MA: Blackwell Publishers, pp. 328–339.

United Nations. 2007. Full Text of the Convention, United Nations Framework Convention on Climate Change. http://unfccc.int/essential_background/convention/background/items/1349.php)

United Nations Conference on Environment and Development. 1992/USA 2002. Explanation of Position by the United States of America Submitted for the Record for Inclusion in the Report of the Conference of the World Summit on Sustainable Development. Released 4 September 2002.

Vaillancourt, Jean G. 2000. Sustainability and Agenda 21, in *Sustainability and Unsustainability on the Road from Rio*, presented at Research Committee 24 of the International Sociological Association conference, Rio de Janeiro, August 2000.

World Bank. 1992. *World Development Report 1992: Development and the Environment*. Washington, DC: World Bank.

Yum! Brands Co. 2007. "About Yum! Brands." Online at http://www.yum.com/about/default.asp.

Notes

1. Vaillancourt (2000) argues that the concept was not new. In 1976 an ecological manifesto by British Greens spoke at length about the need to establish sustainable society; in 1973 and 1977 Ignacy Sacks from France promoted 'ecodevelopment' in the Third World. His 1980 book on strategies of ecodevelopment described ways to 'harmonize ecology and economics'. The WWF, IUCN, UNESCO, and FAO in 1980 put out a report calling for long-term costs and benefits to be included in planning. In 1980 Lester Brown published *Building a Sustainable Society*, and there were many other roots of this idea. (Vaillancourt 2000).

2. Most discussions of this passage emphasize the phrase 'dangerous anthropogenic interference with the climate system'.

3. Some of this discussion is summarized in Roberts and Thanos, 2003.

The State, International Relations and the Environment

Hugh Dyer

The politics of the environment clearly implicate both practices and understandings of the state and international relations. However, the overall question is whether the state and international relations can inform the theory and practice of environmental politics or if in fact it is the reverse; that practices of environmental politics have in some measure transformed the state and international relations, and with this our understandings of these categories. As the discussion below progresses, this question is taken further in considering—even as the topic is discussed in such terms—whether the state and international relations are appropriate categories of thought and practice, either in relation to the environment in particular, or indeed in general. This has implications for the academic disciplinary study of international relations, which may already be feeling the influence of a number of social developments, including the politics of the environment.

THE ENVIRONMENT AND THE LIMITS OF THE STATE

An obvious issue is the limitations of the state as a mechanism for delivering environmental policy. It is difficult for national governments to respond to global environmental change, since their existing political obligations, or perceptions of these, are constrained by prevalent expectations and assumed national priorities. Furthermore, the various ways in which established administrative structures of the state engage with the relatively novel politics of the environment make it difficult to discern the logic of environmental policy output of governments. Perhaps this is, in part, because of the perverse logic of efficiency, rather than the logic of sufficiency, such that apparently successful policies based on efficiency ratios may actually increase overall consumption and ecological degradation in the absence of limits, or a sense of what is 'enough' (Princen 2005). However, there is also a deeper underpinning that makes the environment a difficult subject for the state, its outlook being not just economistic, but (not surprisingly) anthropocentric—meeting human needs and aspirations as judged in isolation from their ecological context. In *Environment and the Nation State*, Lieferrink argues that increased ecological interdependence challenges states' ability 'to control not actually the borders but rather the quality or what may be called the "ecological sustenance base" or "eco-capacity" of their territories' (Liefferink 1996: 26). It is this ecological perspective which raises difficulties for the economistic and anthropocentric habits of the state and international relations and concomitant political and economic practices.

There are a number of key issues to consider: the division of the planet into sovereign states does not reflect the interdependencies of ecosystems crossing state borders. This is important in respect of transnational relations between multiple actors. The state, as a legal entity, is responsible for its own jurisdiction, but can also be held responsible for pollution beyond its borders. State policies adjust to pressure from lobbying groups, but are also subject to other domestic and international pressures. It may be widely assumed by their citizen publics that environmental problems will be managed by governments: 'Most of the strategies put forward for responding to global environmental problems assume that states are willing and able to assume this managerial role' (Lipschutz and Conca 1993: 19). Is this a reasonable or realistic assumption? Given the 'very prevalent suspicion of the state on the part of many ecologists' (Hurrell 2006: 166), is it even desirable? It may be difficult for states to reconcile the different aspects of their responsibilities, and that this is not a manageable situation for states.

There is a growing web of economic, cultural, social, and political relations between states and between states and other actors. While this is clearly a changing political context, the nation state is not likely to disappear (indeed, there are more now than ever before as a consequence of demands for independence). This leaves us to conclude that for the time being at least, non-environmental considerations and domestic political interests will often over-ride commitments to environmental co-operation, and as long as the key environmental agreements are negotiated by states, national interest will play a significant part in decisions made. Nevertheless, this still allows some consideration of what political interests or the 'national interest' are, given that most of the social and political assumptions of national life have been challenged by environmental issues. Barry and Eckersley (2005: 261–263) point to the tension between accumulation and legitimation functions of the state. The former function is supported by weaker win-win versions of ecological modernization that support the globally competitive position of the state (and supply side concern with efficient production), while increasing pressure on the latter function arising from expectations of higher environmental standards (and demand side concern with consumption) points to a stronger version of ecological modernization which implies a need for more clearly transnational political and economic practices. There may also be more fundamental doubts about whether and which interests are served by the technological optimism, reformism, and 'statism' of ecological modernization (Mol 2001; Fisher and Freudenburg 2001; York and Rose 2003). The state may remain, but its context and position may change.

The traditional political goals of society which the state purports to serve— such as health, wealth and security—are likely to be viewed differently in an environmental light, requiring development of sustainable policies even if within the constraints of existing social and political systems in the first instance. However, achieving any measure of success in environmental policy is likely to require substantial change in habitual political practices of decision-making and agenda-setting, and the means of wealth creation and protection of national interests. Equally, or in parallel, there are challenges to existing social practices, such as uneven distribution of resources, the character of the capitalist economic system in respect of profit motive and pressures to increase productivity, and the corollary of such economic growth—increased and increasing consumption. In many respects, therefore, both state and society may be wrestling with a set of conflicting goals, though this might yet be addressed if there remains scope for reframing and even reconstructing identities and interests.

To exacerbate this situation, there are various constraints on social and political change, at individual, institutional, and international levels. Constraints at the level of individuals exist in the embedded assumptions and habits of individuals and in the attitudes of the general public to environmental issues—these may change, of course, but perhaps not very quickly. The existing version of the capitalist system serves and supplies individual 'needs', and so reinforces and is reinforced by individual and public attitudes about the appropriateness of economic growth, individualism, competition and self-interest. At the next, institutional level, there are constraints in that the attitudes of individuals are rooted in political, social and economic institutions which are not designed or developed to implement sustainable goals. At the international level, the con-straints relate to the authority of the state in decision-making processes, economic competition between states, and the relative weakness of international regimes.

Eckersley, in *The Green State*, points to three core challenges for the prospect of a green state: anarchy in the states system; promotion of capitalist accumula-tion; democratic deficits in the liberal state. She argues that the key to transfor-mation is increased accountability to both global civil society (citizens and others) and international society (state-based organizations and institutions). The logic of this structure, she allows, may be challenged by the emerging environmental multilateralism, sustainable development strategies, and environmental advocacy, though crucially the success of such a challenge is dependent on a distinctly green conception of state governance (Eckersley 2004: 14–15). We can see some evi-dence of this tension between accountability and the pursuit of capital accumula-tion in recent attempts to establish more ambitious carbon reduction targets coinciding with liberalization of transatlantic air travel (with increased competi-tion likely leading to increased flights). We might also consider that the world's merchant fleet, essential to world trade, is so far exempt from carbon emission limits; even the recent and potentially radical British legislative proposal for legally binding five-year limits on carbon emissions excludes aviation and shipping initially (*The Times*, 14 March 2007). Finally, in the context of climate change, time is not on our side and the importance of longer-term transformation may be displaced by the urgency of shorter-term action—such is the difficulty of escaping the immediate logic of established political and economic practices in order to adopt a more ecological perspective.

The existing structure of both the international legal and political systems rests heavily on independence and autonomy of states, and so collective environmental management poses politically sensitive challenges, involving the creation of rules and institutions which reflect the rather different idea of shared responsibilities. There may also be a range of apparently reasonable grounds for resistance by states to an environmental supranational authority: the state remains a source of human identity, and is a significant means of political expression, which gives claims to sovereignty moral credibility. The significance of environmental challenges, though important, may not be sufficient reason to abolish sovereignty when it is anyway not clear that supranational authority would lead to efficient environmental management. Litfin has indicated that experience of environmental regimes 'warrants a healthy scepticism about whether the nation-state system can smoothly adapt to ecological interdependence via traditional forms of multilateral, state-centric institutions' (Litfin 1993: 111). However, there may be other reasons to abandon strict versions of sovereignty, including the absence of any choice about the matter. Under the heading 'Sovereignty and the inadequate state', Elliott argues that it is:

> 'not simply that the unilateral state cannot meet the challenges of global environmental change through self-help when the causes of that change lie outside its borders. It is that the state itself—its autonomy, capacity and legitimacy—is being eroded, or at least challenged, by the very nature of environmental problems which do not respect territorial borders.' (Elliott 2004: 109)

Even so, the international states system per se is only one factor in the management of the global environment, and there are signs of change in domestic polities in terms of policies on pollution and waste management (via taxation or regulation) and increased environmental awareness at least in the limited terms of managing environmental risk. Individuals have undergone changes to attitudes and practices, with consumer activism in the 'North' and producer activism in the 'South', and are increasing their political leverage through public demand for increased transparency and involvement in policy formulation. Even those aspects of the international system which are able to escape a purely state-centred perspective may influence the behaviour of states and other actors, and international engagement can both help to promote domestic policy goals as well as underwrite international law.

If the changing position of the state can be attributed to public demand rather than governmental initiative, then we should perhaps consider the importance of public opinion in the creation and formulation of state policies for the environment, and equally how public pressure plays a similar role at the global level. Of course, public opinion is difficult to assess, and there is also a question as to what constitutes a relevant 'public' (if we can't assume that this is already constituted by the state in terms of citizenship and electoral registers). It would be problematic to assume that 'the public' is constituted by unelected élites, or unrepresentative activists. Nevertheless, there is plenty of evidence for public opinion carrying weight on environmental issues, if it is often difficult to distinguish from the influence of organized non-state actors. A good example is the change led by influential NGOs mobilizing world public opinion behind the rights of whales.

Other familiar examples include seal culling, disposal of the Brent Spar oil storage container, French nuclear testing in the South Pacific, etc. In such cases, governments are required to balance 'national interest' with the need for public support. Such cases of public opinion driven by environmental activism have normative significance, and reflect 'universalistic moral concern and a conception of collective human interest' (Vogler 1995: 201). Vogler suggests that it is easy to be cynical about moral positions in politics, but public support for such positions cannot be discounted. We could even consider extending the moral community, and indeed some form of representation, to the non-human realm in an ecocentric approach, notwithstanding the challenges of integrating such positions into current practice (Eckersley 1992).

There is further evidence of this trend in non-state politics in the emergence and relative significance of environmental social movements. Dryzek's volume on Green States and Social Movements suggests that social movements are influenced by the kind of state they relate to, and conversely states may be transformed by incorporation of, or resistance to, social movements. This has implications for the choice to work on the state, or to work around it through civil society, as political strategies for environmental movements (Dryzek, 2003). We might conclude that if the environmental situation is a cause of political behaviour, and is not improving, then the drive towards environmentalism is one which states may not be able to resist.

INTERNATIONAL RELATIONS AND THE ENVIRONMENT

The state remains at the centre of debate. Dauvergne discusses the possibility of a 'secure world of states, institutions and regimes' with circumspection. He notes that some arguments assume that global institutions and regimes cannot constrain the self-interested behaviour of states (that damages common resources), while other arguments support global governance on the basis of complex drivers and constraints on states, and the rational choice of states to co-operate through management regimes and develop institutions (Dauvergne 2005: 13–16). Consequently, the focus of critique in respect of disciplinary international relations (IR) is the conventional preoccupation with the state as a constitutive central actor in the practice of international relations. In particular the traditional attribute of the state, sovereignty, is seen as a constitutive concept of international relations. Karen Litfin's edited volume, *The Greening of Sovereignty in World Politics*, engages this issue with a collection of stimulating essays that examine the less-than-obvious relationship between sovereignty and ecology, countering the commonplace assumption that the two are irreconcilable by pointing to ways in which sovereignty is revisioned, reoriented, and problematized in respect of the relevant socio-political boundaries and concepts. Specifically, she argues that 'conceiving of sovereignty in terms of autonomy, control, and authority usefully decenters the state', and that sovereignty 'can be an attribute of various political entities, not just the state' (Litfin 1998, 9).

Conca, in his study of the relationship between extremes of interstate regimes and global civil society, sets out two challenging observations on the role of the state, while acknowledging that the state does have some role to play. The first challenge is a poor track record of centralization and industrialization at the expense of the environment. The second challenge is that globalization has cost the state some potential ability to respond to environmental issues, even if states were/are complicit in the deregulation and liberalization of a transnationalized world economy (Conca 2005: 181–182). Globalization is uneven and hierarchical, and the competitive aspects of international relations remain, supported by notions of relative gain in a zero-sum situation of scarcity. However, in the context of global environmental change absolute gains are more likely in the long term, even if immediate costs imply relative gains in the short term. So in this respect the stakes are high for those state actors considering political and economic integration. For those inclined to protect their borders and economies from the effects of globalization (and environmental change), this approach to protecting sovereignty may come at a high price.

If environmental governance is tethered to broader processes of globalization and associated forms of global governance, and if these can draw attention away from state-centric concerns towards the global/local, then the question is whether globalization can be good for the environment. If it could be, could it also be good for people? This may still be an open question; but even if the anti-globalization movement provides an obvious case of resistance this is focused on exploitation and inequality rather than constructive global co-operation. Clearly, if there is any emerging consensus on a new global environmental order, then one or more alternative existing order would be displaced by it. Perhaps the existing orders are 'development' (but this means different things to different peoples) and 'sovereignty' (but this has always been elusive), and there are those who would be happy to see the end of some versions of either or both of these. For example, 'environment and development' has become an established (if unclear) agenda of international relations for the 21st century. That this agenda reflects earlier agendas of colonization, decolonization, and uneven development (longstanding 'North–South' issues) perhaps reduces its novelty, but it adds environmental concern—even if disguising the political interests of the North. This agenda points to the complexity of environmental issues and their interdependence with other functional issue-areas, all set against the economic and political challenges of countries pursuing industrial growth. It also raises again the question of what (not to mention if) global institutional arrangements might ameliorate the situation. Young argues for a common research agenda or dialogue in order to develop a unified theory of environmental governance, in spite or because of divergent research foci—representing bottom-up and top-down perspectives—on (a) cases of local common property systems in small-scale settings (Young 2005: 176) and (b) disciplinary debates that seem 'more concerned with conceptual and methodological matters than with advancing understanding of major substantive issues like identifying the conditions under which environmental regimes will produce outcomes that fulfil various criteria of sustainability, efficiency or equity'

(Young 2005: 178). A useful illustration is George Monbiot's recent attack on biofuels, arguing (as do other critics) that the claimed benefit of reducing carbon emissions (as carbon is sequestered in plant growth) is undone by the consequences of the global market for these fuels leading to carbon-inefficient processes (indeed net carbon increases that exceed those of fossil fuels) and deforestation and displacement of food crops in places where they are most needed (*The Guardian*, 27 March 2007). Some carbon offsets have a similarly double-edged quality, as do electric- or hydrogen-powered vehicles which merely displace emissions, and without market incentives provided by global regimes the 'price of carbon' may not be sufficient to deliver appropriate technologies widely. This suggests that technological optimism is unwarranted, and that rather more subtle, complex, and systemic solutions may be needed to square the circle of environment and development. This also suggests a case for humility in the face of such political challenges.

With any global political scheme comes the danger of substituting one over-arching discourse for another, and this may simply replicate hegemonic or imperial tendencies. In the current political imagination, global environmental governance is constrained by, or aligned with, the fashion for deregulation and liberalization such that win-win solutions (which may have merit if winning is ecological) are promoted under the banner of economic efficiency. Perhaps eco-capitalism is the appropriate charge against current global environmental governance, rather than eco-imperialism. Even so, assuming the hegemonic aspects of 'primitive accumulation' remains problematic in a world where social transformation is so varied (Shilliam 2004). Lipschutz argues that 'it is the relationships between ruler and ruled, and the mechanisms of rule, that are important' and cites two models of empire: 'neo-liberal institutionalism' and 'new sovereignty' (Lipschutz 2004: 21). Here again, we may ask if global environmental policy represents a new form of imperial governance to be resisted, or a gradual transformation in the mechanisms of authority and legitimacy of states—that is, something that might be welcomed.

Conca argues that at the same time as states are displaced in some degree by transnational civil society, there remains some exercise of state authority (Conca 2005: 183). However, this occurs in a rather different context of institutionalized politics, in which the state is not irrelevant but its authority is contested. (Conca 2005: 194–196). He concludes that governance is increasingly transnational, institutions are more complex, and exercise of authority is more fluid (Conca 2005: 202–203). Weiss and Jacobson (1998) point out aspects of relationships between individual and collective actors in the international system which should tell us something about the way international relations is changing—and how this might change our thinking about it. Perhaps not surprisingly, it turns out that actual implementation of environmental agreements is the greatest challenge. In order to judge the success of environmental diplomacy it may be necessary to know how and why implementation and compliance with such agreements varies. Interestingly, the international political environment remains a dominant factor, and not least because while states are obviously central to implementation and compliance none can or will act in a vacuum. The term 'engagement' goes some way to capturing the political dynamic, and is Weiss and Jacobson's prescription for future compliance. Compliance is typically both a legal and technical issue, but ultimately behaviour modification is what counts, and in this respect non-governmental actors and communities of technical experts make considerable contributions to what may be seen as both transnational and transgovernmental activity (Vogler 2000). Stiles takes a pluralist view of the relationship between civil society actors and states and intergovernmental institutions, and argues that 'the interests and identities of major players tend not to change over time, only their strategies and tactics depending on the general distribution of power and resources' (Stiles 1998).

This pluralist perspective on the state system, and its relation to other actors, may be both accurate and politically appealing: 'it is important to note the normative claims made for this kind of [classical] pluralism' (Hurrell 2006: 166). Yet the pluralist view, and practice, may be fundamentally (if unintentionally) reactionary: 'Indeed, green arguments that economies should be brought back under firm national control and the 'excessive' immigration should be resisted attest to the continued power of the pluralist impulse' (Hurrell 2006: 167). This seems to make sense in terms of contemporary practices, but Smith (1993) warned against an unchallenged pluralist consensus and absence of more critical engagement as being a cause of the environment's marginal place in IR—though of course that may say more about IR than about environmental politics.

Brenton, in considering the role of the state, suggests some caution in regard to supranational rather than subnational, local, individual approaches: 'I would not want to conclude with the impression that solving the world's environmental problems is finally a matter of creating the right international institutions and rules.' He suggests that the collective and integrative perspective of environmentalism has made it easy to accept 'dense webs of regulation and grand schemes of national and international planning (sometimes, as we have seen, with the suppression of individual liberties as an accompaniment)'. Clearly some goals can only be achieved by international co-operation given the global environmental and political-economic context, but the modes and consequences of such co-operation remain an issue. Reflecting widely held concern about the nature of political authority and accountability in the supranational context, Brenton goes on to note that 'replacement of the judgement of the individual by that of the state raises problems of its own', and this is compounded by transfer of authority to the international level 'placing it still further from the people it is intended to serve...' (Brenton 1994: 268).

THE INDIVIDUAL AND THE ENVIRONMENT

There are of course good ethical and political reasons for privileging the individual over the state or any other political authority: on the simple premise that 'Each of us has an overriding obligation to be morally autonomous; ... the conclusion was quite outrageous: A morally legitimate state is a logical impossibility'

(Wolff 1998: vii). Individuals may well hold the secret to dealing with the environmental crisis, either as challengers of technocratic society (Roszak 1979), or as a source of ethical and political meaning (Peterson 2001). Yet if we hold the individual to be fundamental in ethical and political terms, we are nevertheless obliged to consider the wider moral context, and what these 'political terms' are. It may be that the conventional political location of individuals within the defining purview of the state is inadequate for the purposes of environmental politics, and as Beitz (1979: 180), for example, notes, 'the critique of the idea of state autonomy clears the way for the formulation of a more satisfactory normative international political theory', though he elsewhere notes that political theory should guide rather than replace practical judgement (Beitz 1989: 227). We might observe that, for the most part, people are eminently (and immediately) practical in ways that political institutions are often not.

The individual, even in the context of world environmental politics, is a significant political fact—clearly so in an individualistic ontology of politics (as in conventional liberalism, for example) but this would be difficult to avoid for all but the most objectionable (totalitarian) political perspectives. What is more, the individual has a convincing claim to be both source and content of value, and so it is only by finding a place for the individual in world environmental politics that we can determine the source and content of the relevant political values. While a more holistic ecocentric perspective would certainly challenge the atomistic and fragmented account provided by individualism, ecocentric values will need to be held politically and the individual may still be the key to finding a political home for such values given that the state is less amenable to them.

Out of this emerges perhaps only an attitude towards, or a story about, world environmental politics in which the individual is implicated as either the source or locus of political values, or both. Even if ecocentric in its intentions it remains a weakly anthropocentric story, partly as a consequence of the human subject, but also because politics is an anthropocentric exercise in which such story-telling is important. Peterson, in examining Oelschlaeger's writing (in *Caring for Creation*, 1994), identifies his suggestion that humans are (in the context of socially created ecocrisis) 'storytelling culture dwellers', in contrast to received assumptions about rational self-interested agents (Peterson 2001: 8). A focus on the individual in turn has implications for theorizing world environmental politics as a particular facet of international studies that increasingly commands attention, and in doing so reorients our understanding of international relations or global politics more generally. Specifically, from the centrality of the individual it follows that neither cosmopolitan nor communitarian references provide entirely satisfactory moral foundations for world environmental politics, and that universalizing discourses typical of modernity are also not suitable points of reference even if translated into the holistic ecological idiom.

The political implications of this perspective may be rather more trans-social than interstate, more global/local than international, while allowing that the individual retains moral and political standing. Accepting the individual necessarily requires a more contingent and subjective account of world environmental politics, and so the challenge is to understand the contingent circumstances. This still leaves us with some room for considering the social influences of environmental politics.

THE SOCIAL AND THE ENVIRONMENTAL

Kütting (2000) connects environmental degradation to social origins, and shows that the failure to recognize the centrality and complexity of this connection has resulted in its externalization through concentration on the study of international institutional developments. To the extent that the literature tends to reflect scientistic rational analysis which largely ignores such underlying social issues, the implications for the study of international relations are quite broad, and in particular point less to the 'international' and more to the 'social'.

If we cannot evade the political, we can surely see that the institutional and the social are implicated in our understanding of global politics. Elliott's admirable survey of The Global Politics of the Environment begins by noting two 'simple aphorisms'—that 'global environmental problems ... require global solutions', and that there are 'no simple solutions'—and puzzles about 'how we should understand the "global" as an organising principle' (Elliott 2004: 3–4). We are perhaps most concerned about the direction of political causation (such positivist social science pursuits die hard), perhaps for good practical and ethical reasons, and in this lie both potential for change and potential dangers. The case of 'global governance' is a useful test, describing something short of government in its state-centric sense, and indicating formal and informal structures and processes, all in aid of (potentially global) political order; but it remains a form of 'politics from above' (Maiguashca 2003: 5). In its global manifestation this may well amount to mere hegemony or imperialism if global environmental governance is understood as a device for protecting existing power structures rather than changing them—a 'globalisation from above' (Elliott 2004: 111–112). In the face of any such global governance initiative we could expect to see (and have seen) some signs of resistance by individual activists, social movements, non-state actors, and perhaps weak state actors, representing a form of 'politics from below' (Maiguashca 2003: 5). This hierarchical 'above' or 'below' seems typically political, when governance issues may be better understood as social. In any case the environmental crisis adds a particular additional consideration to ethical, social, political and economic (anthropocentric) tests we might apply to any scheme—that of ecological integrity.

Schemes to improve the human condition do not always work out according to plan, even if it is possible to point to progress in some respects. There is no reason to think that we will cope much better with environmental problems than we have with problems of inequality, not least because they are linked. However, there may yet be some progress over the long term. In years to come, there may be a call for expressions of regret for contemporary environmental practice in much the same way as there are presently such calls for apologies to mark the bicentenary of the Abolition of the Slave Trade Act. The comparison is useful, not because expressions of regret will right a wrong (or that there is moral equivalence between the two cases), but because it illustrates a constraint on behaviour that was widely unaccepted over 200 years ago, the absence of which is widely unthinkable now. It also illustrates how changes in authoritative values and practices, and the nature of struggles to change practices and values (slave revolts preceding legislative abolition), are not linear or uniform or complete (enslavement takes different forms over time; practices were and are uneven across space; types of slavery still exist). It also illustrates the tension between economic advantage and the proper exercise of political (or moral) authority, even if the dominant social paradigm of the day expresses a recognizable consensus on the issue.

The case of the environment may follow such a pattern of social change, in the context of its own times, such that what may be viewed as unrealistically burdensome constraints on behaviour are in the future seen to be clear moral requirements and become both commonplace and common sense. What is more, facing the environmental challenge need not be seen in negative terms of constraint, but can be readily understood in positive terms of opportunity. Princen has convincingly shown that sufficiency, rather than 'efficiency', is an entirely practical goal, which results not in merely surviving but in actually thriving (Princen 2005: 3). The challenge for people and politics is to underwrite such opportunities as being legitimate. The nature of the illegitimacy (or political distance) is illustrated by Perkins in indicating that where the political scale of decision-making (e.g. the state) is at odds with the ecological scale of environmental impacts this results in a democratic deficit (Perkins 2000). Thus the consequence of the environmental crisis for the state and international relations is that it 'calls into question both the practical viability and the moral adequacy of this pluralist conception of a state-based global order' and this has already elicited a partial response in that it has 'pushed states towards new forms of international law and global governance' (Hurrell 2006: 167), even if it has not yet brought about fundamental change in the world's social structure.

CONCLUSIONS

The concerns discussed above require us to consider—in looking at the state, international relations and the environment, and the study of these in disciplinary IR—the radically alternative case for not looking at the state, international relations and the environment, and the study of these in disciplinary IR, in the sense that we should 'Forget IR Theory' (Bleiker 1997). If the state is both resilient and recalcitrant, the terms state and international relations (and IR) are still a way of policing a particular framework of thought and action, which in the end cannot account for much of what is significant in politics, including the environment. For example, Weber (1999) argues that apparent disciplinary innovations in IR such as constructivism simply incorporate and assimilate radical critiques by evacuating their political significance through exclusion of 'everydayness' in the disciplinarity of IR. She supports 'forgetting the object of critique', citing Bleiker (1997).

Thus the state and international relations represent a set of political structures, with roots in self-determination and independence, rather than a set of tools for dealing with environmental issues. If such tools cannot be derived from these political structures, concern with the role of the state and international relations becomes less relevant, and concern with human practices and the role of political economy and transnational civil society becomes increasingly relevant. What is more, since the concern remains anthropocentric rather than ecocentric, a shift in analytical relevance away from state-centric politics is not the end of the story from an environmental perspective. In this sense, difficulties with the recalcitrance of the state may simply be reproduced in the recalcitrance of politics in other guises, or in other modes of human organization. If the recalcitrance of politics was a *raison d'être* of the state, then it may be felt that escaping state-centric structures is merely displacing the difficulty—if we did not have such administrative structures we would have to invent them in order to deploy collective resources, or something similar but less driven by requirements of technological rationalism and hierarchical authority. Nevertheless, shifting our gaze away from the state as an object of analysis will allow a more interdisciplinary view of the political subject matter, or at least a more nuanced notion of the state as an organ of human interests.

For example, George Lawson's defence of historical sociology as an approach to international relations claims that its 'rejection of universal, timeless categories and their replacement by multilinear theories of world historical development give history the chance to breathe and agency the chance to make a difference'. As well as allowing for 'periodicity' (in the sense of understanding of the emergence of the environmental), this may allow us to acknowledge the emergence of 'an epoch defined primarily by globalization and deterritorialization' that more readily imagines 'the importance of time and place variation' (Lawson 2006: 415–416). He allows that there are many starting points for theoretical concerns that reflect such an historical sociological perspective, including the emergence of global norms across temporal and spacial locations, and here we can see room for exploring environmental norms that underpin world environmental politics and the role of individuals in their development.

Given the complexity and ubiquity of environmental issues, theoretical guidance is in great demand for expedient reasons of description and explanation, as well as the need for greater understanding. These increased opportunities for theory provide environmental concern with potential new avenues of expression and application, and with environmental values running through modern political discourse (Dyer 1996), there is the possibility for constructively subversive developments in the traditional disciplines. Where disciplinary concerns in international relations can be (and often have been) articulated as the calculus of interests leading to either conflict or co-operation, refocusing attention on values might point to alternative characterizations of the political dynamics. When 'rational interests' do not have to be satisfied, in order to make sense of international relations, then the theoretical possibilities are open (some notion of interests may remain, of course, but need not be determinate). One illustration is Tim Hayward's *Political Theory and Ecological Values*, arguing that environmental values can be supported by enlightened human interests, that this link must exist if ecological goods are to be promoted, and that there are profound implications of fully integrating environmental issues into our disciplinary concerns (Hayward 1998). If international relations scholars are concerned with the transformation of political community (Linklater 1998) rather than the preservation of the environment, once environmental issues are introduced, the fundamental problematic becomes the transformation of the human relationship to the environment. So a possible consequence of engaging with the environmental is that international relations are less about conflict and co-operation as solutions to pursuits of state interest, but perhaps, as I have argued elsewhere (Dyer 2000), more about coping with competing values and the practical means of dealing with them. In this sense we are already concerned more with 'global politics' than 'international relations', though it is not clear if this amounts to a transformation.

As Hobson and Hobden put it, in respect of one of their proposed research agendas for world sociology, we are obliged 'to rethink the origins of international systems, states and international institutions as well as to denaturalize such historical forms, and to consider the potential and actual processes which are reconstituting, if not transforming, the present into possible and desirable futures' (Hobson and Hobden 2002: 283). Bigo and Walker observe that a sociological approach to politics and international relations offers the benefit of 'emphasis on the study of practices' including discourses, rather than lapsing into engagement 'with systems, states, sovereignties and so on as more or less disembodied structures, even abstractions' (Bigo and Walker 2007: 5). Environmental change exacerbates the situation of states by creating different contexts which are not amenable to state-centric approaches, but for which states are at the same time held (or rather held to be) responsible. In this latter respect both everyday demands for state action on the environment (or, as often, resistance to it), as well as disciplinary IR's reification of the state as the locus of politics, create obstacles to transformation. Nevertheless, what might now better be called 'world environmental politics' dislodges conventional 'international' or even 'global' understandings of moral and political agency. It does so by virtue of its transnational character and the increasing significance of non-state contexts, factors, as actors such as civil society and NGOs, the market and transnational corporations, the environment itself (as either context, factor, or 'actor'), and finally the lowly individual (as citizen, activist, consumer, producer, or moral and political agent). In this wider socio-political-economic context, ecological significance may be the determining factor in the end.

It can be argued that environmental concern has already had a profound influence on world politics, in practice and in theory, and we may begin to view the traditional staples of the field—state, nation, international relations—as artefacts of previous attempts to resolve the old political questions. These may be resolved differently in the ecological context of a world environmental politics which cuts across political, social and economic boundaries. However resilient and recalcitrant as a political form, the state is left in an awkward position, and may never be quite the same again.

Bibliography

Barry, J., and R. Eckersley. 2005. W(h)ither the Green State?, in *The State and the Global Ecological Crisis*, edited by J. Barry and R. Eckersley, 255–272. Cambridge, MA: The MIT Press.

Beitz, C. 1989. *Political Equality: An Essay in Democratic Theory.* Princeton: Princeton University Press.

Beitz, C. 1979. *Political Theory and International Relations.* Princeton: Princeton University Press.

Bigo, D., and R. B. J. Walker. 2007. Editorial. *International Political Sociology.* 1 (1): 5.

Bleiker, R. 1997. Forget IR Theory. *Alternatives.* 22 (1): 57–85.

Brenton, T. 1994. *The Greening of Machiavelli.* London: Earthscan.

Commoner, B. 1971. *The Closing Circle: Nature, Man and Technology.* New York: Knopf.

Conca, K. 2005. Old States in New Bottles? The Hybridization of Authority in Global Environmental Governance. In *The State and the Global Ecological Crisis*, edited by J. Barry and R. Eckersley, 181–205. Cambridge, MA: The MIT Press.

Dauvergne, P. (Ed.). 2005. *Handbook of Global Environmental Politics.* Cheltenham: Edward Elgar.

Dyer, H. C. 2000. Coping and Conformity in International Relations: Environmental Values in the Post-Cold War World. *Journal of International Relations and Development.* 3 (1): 6–23.

Dyer, H. C. 1996. Environmental security as a universal value: implications for international theory. In *The Environment and International Relations*, edited by J. Vogler and M. F. Imber, 22–40. London: Routledge.

Eckersley, R. 2004. *The Green State: rethinking democracy and sovereignty.* Cambridge, MA: The MIT Press.

Eckersley, R. 1992. *Environmentalism and Political Theory: Towards an Ecocentric Approach.* London: UCL Press.

Elliott, L. M. 2004. *The Global Politics of the Environment* (2nd edn). Houndmills: Palgrave.

Fisher, D. R., and W. R. Freudenburg. 2001. Ecological Modernization and Its Critics: Assessing the Past and Looking Toward the Future. *Society and Natural Resources.* 14: 701–709.

The Guardian. 27 March 2007.

Hayward, T. 1998. *Political Theory and Ecological Values.* Cambridge: Polity Press.

Hobson, J. M., and S. Hobden. 2002. On the Road Towards an Historicised World Sociology. In *Historical Sociology of International Relations*, edited by S. Hobden and J. M. Hobson, 265–285. Cambridge: Cambridge University Press.

Hurrell, A. 2006. The State. In *Political Theory and the Ecological Challenge*, edited by A. Dobson and R. Eckersley, 165–182. Cambridge: Cambridge University Press.

Kütting, G. 2000. *Environment, Society and International Relations: Towards more effective international environmental agreements.* London: Routledge.

Lawson, G. 2006. The Promise of Historical Sociology in International Relations. *International Studies Review.* 8 (3): 397–423.

Liefferink, D. 1996. *Environment and the Nation State.* Manchester: Manchester University Press.

Linklater, A. 1998. *The Transformation of Political Community.* Cambridge: Polity Press.

Lipschutz, R. D., and K. Conca. 1993. Act I: The State and Global Ecological Interdependence, in *The State and Social Power in Global Environmental Politics*, edited by R. D. Lipschutz and K. Conca, 19–23. New York: Columbia University Press.

Lipschutz, R. 2004. Imitations of Empire. *Global Environmental Politics.* 4 (2): 20–23.

Litfin, K. 1998. The Greening of Sovereignty: An Introduction, in *The Greening of Sovereignty in World Politics*, edited by K. Litfin. Cambridge, MA: The MIT Press.

Litfin, K. 1993. Ecoregimes: Playing Tug of War with the Nation-State. In *The State and Social Power in Global Environmental Politics*, edited by R. D. Lipschutz and K. Conca, 94–117. New York: Columbia University Press.

Maiguashca, B. 2003. Governance and Resistance in World Politics: Introduction. *Review of International Studies.* 29 (1): 3–28.

Mol, A. P. J. 2001. *Globalization and Environmental Reform: The Ecological Modernization of the Global Economy.* Cambridge, MA: The MIT Press.

Perkins, E. 2000. Equity, Economic Scale, and the Role of Exchange in a Sustainable Economy, in *Nature, Production, Power*, edited by R. M. M'Gonigle and F. P. Gale,183–194. Cheltenham: Edward Elgar.

Peterson, A. L. 2001. *Being Human: Ethics, Environment and our Place in the World.* Berkeley: University of California Press.

Princen, T. 2005. *The Logic of Sufficiency.* Cambridge, MA: The MIT Press.

Roszak, T. 1979. *Person/Planet: The Creative Disintegration of Industrial Society*. London: Gollancz.

Shilliam, R. 2004. Hegemony and the Unfashionable Problematic of "Primitive Accumulation". *Millennium: Journal of International Studies*. 33 (1): 59–88.

Smith, Steve. 1993. The Environment on the Periphery of International Relations: An Explanation. *Environmental Politics*. 2 (4): 28–45.

Stiles, K. 1998. A Rational Choice Model of Grassroots Empowerment. International Studies Association paper, http://www.ciaonet.org/conf/stk01/ viewed 27/3/07.

The Times, 14 March 2007.

Vogler, J. 2000. *The Global Commons: Environmental and Technological Governance*. Chichester: John Wiley & Sons.

1995. *The Global Commons: A Regime Analysis*. Chichester: John Wiley & Sons.

Weber, C. 1999. IR: The Resurrection or New Frontiers of Incorporation. *European Journal of International Relations*. 5 (4): 435–450.

Weiss, E. B., and H. K. Jacobson (Eds). 1998. *Engaging Countries: Strengthening Compliance with International Environmental Accords*. Cambridge, MA, and London: The MIT Press.

Wolff, R. 1998. *In Defence of Anarchy*. Berkeley: University of California Press.

York, R., and E. A. Rose. 2003. Key Challenges to Ecological Modernization Theory. *Organization and Environment*. 16 (3): 273–288.

Young, O. R. 2005. Why is there no unified theory of environmental governance?, in *Handbook of Global Environmental Politics*, edited by P. Dauvergne. Cheltenham: Edward Elgar.

International Political Economy and the Environment

MATTHEW PATERSON

INTRODUCTION

The political dynamics of the global economy are central to determining both responses to specific environmental challenges, and the possibility of pursuing sustainability more broadly. Does the global expansion of the economy necessarily contradict the principle of sustainability, or can the economy be transformed so that it continues to grow while reducing its environmental impacts dramatically? Does the power of large multinational firms, or of global institutions like the World Bank, act to prevent action on climate change, or promote the generation of environmental hazards such as in the toxic waste trade? These are the sorts of questions that lead us to consider International Political Economy (IPE), the field of study that examines questions such as the power of global corporations, the dynamics of global economic expansion, or the role of multilateral economic institutions.

This essay elaborates four key themes and questions that arise from the point of view of IPE in relation to environmental politics. These are: the interaction between the way the global economy is managed collectively by states and the environment—for example the interactions between trade, its environmental impacts, the tensions which can arise between the trade regime and environmental regimes; the power of multinational firms in their relations with states and social movements; the question of economic growth—whether it has basic environmental limits or not, and how to deal with the differences in wealth across the globe in the context of potential ecological limits; and finally different perspectives on the overall compatibility of a 'free market' or 'capitalist' economy with the principle of sustainability. Before we get to this, however, it is worth giving a brief overview of IPE, which provides the core concepts through which we can generate our specific questions concerning environmental politics.

IPE—A SHORT OVERVIEW

Scholars working in IPE would quickly disagree of course on what IPE is. For the purposes of this essay, it is reasonable, however, to identify two basic conceptions. Handily, they map fairly well onto the different themes identified regarding environmental politics. For one set of perspectives, usually known as 'realist' (e.g. Gilpin 1987) and 'liberal' (e.g. Keohane 1984) respectively, the aim of IPE is to explain the key dynamics of the interaction between the international economy and the interstate system. States and markets are understood as different aspects of social life, with their own basic logic and dynamics, but which necessarily interact with each other. But because they have competing logics—states being institutions organized 'vertically' through the imposition of their rule and authority over particular territories, competing with each other for territory, power and influence over other states, while markets are 'horizontal' institutions where individuals meet to pursue their interests, and which in principle do not operate on a territorial basis—there are frequently tensions between the two. Most fundamentally, the pursuit of power and national interests by states frequently gets in the way of the smooth operation of an international market economy on which generalized prosperity is assumed to depend. The basic purpose of IPE is therefore, from this perspective, to analyse whether and under what conditions states can co-operate sufficiently to produce stable international economic expansion, to produce if you like the 'public good' of free trade, international investment, and so on, or in our context, resolution of environmental problems. Realists are significantly more pessimistic about the possibilities of such co-operation than liberals, and also tend to attach more normative value to states than to markets.

For a group of radical perspectives, this conventional separation of the global political economy into 'states and markets' (Strange 1988) is basically mistaken, if not fundamentally ideological. Rather than IPE being about the interaction of two sets of institutions with different and competing logics, the global political economy should be understood as an integrated whole—a capitalist society—in which varying institutions such as states, social movements, firms, and markets, have historically emerged together. Capitalism has always had a globalizing logic, but we live now in a period where its power can be felt on a truly global basis. The goal of IPE is therefore fundamentally to explain the dynamics of capitalist society as it continues its world-wide expansion. The core perspective here is often known in IPE as 'neo-Gramscian', after the use by many scholars of the work of Gramsci (e.g. Gill and Law 1988; van der Pijl 1998), but feminists (Mies 1986) and many Greens (Helleiner 1996; Paterson 1999) writing on IPE would share this basic resistance to the idea that states, firms and markets are basically in tension with each other. Rather, given that capitalist expansion benefits principally capitalists, and states have evolved historically to support the expansion of the capitalist economy, one expects to find in fact significant collaboration between states and firms to pursue the generalized expansion of capitalism, often at the expense of other interests (workers, the environment, for example). This perspective is therefore much more explicitly political and normative in character; the goal of analysis in IPE is not only to identify why the global economy works the way it does, but to identify how the interests of the marginalized or exploited may be pursued in opposition to the power of multi-nationals and powerful states.

From these two perspectives, it is fairly easy to see that the questions and analyses that might be generated when each tries to consider environmental questions will be significantly different. I will try to elaborate these differences as

I proceed below. But broadly, we can surmise for example that realists will be sceptical that the interstate system will allow the widespread co-ordination of the global economy that many suggest is necessary to produce sustainability. Many liberals are more optimistic on this point, but worry on the other hand that environmentalism or responses to specific environmental challenges may compromise the expansion of the free market economy pursued more or less successfully since 1945. Neo-Gramscians and others will on the one hand place emphasis on how the power of global business shapes and mostly limits responses to environmental problems, while also asking questions about the role of environmental movements in opposition to global capitalism.

THE ENVIRONMENT AND THE MANAGEMENT OF THE GLOBAL ECONOMY

Clearly, from both sets of perspectives, how the global economy is organized and managed is crucial from an environmental point of view. As a consequence, there is a large literature in IPE on this theme. The management of the global economy is frequently divided into two principal areas—trade on the one hand, and money and finance on the other. Both areas have become contentious in relation to environmental problems.

Trade is often thought to be the motor of the expansion of the global economy. For conventional economists, it is the pursuit by firms and countries of their comparative advantage which leads to increasing specialization and thus international trade. After the Second World War, the General Agreement on Tariffs and Trade (GATT) was established to try to reduce barriers to trade erected by governments seeking to protect their own economics from international competition. This contributed to an enormous expansion in the volume of international trade in the following decades. The multilateral management of trade, now in the context of the World Trade Organization (WTO), created in 1994, is still dominated by a concern to remove remaining such barriers to the movement of goods.

As soon as people started thinking about environmental questions, however, the benefits of trade were brought into question. There is now therefore a significant debate on the trade-environment connection. It is worth distinguishing two elements here—the debate about whether trade itself makes environmental degradation worse, and the debate about the management of trade by the WTO and whether this impedes measures to improve the environment.

Trade frequently entails shipping goods across vast distances, often goods which could be produced in the importing country. As well as by multilateral co-operation, trade in the post-Second World War period was also stimulated by significant reductions in the costs of travel and increasingly by air transport. This has considerably increased the environmental impacts of trade given the energy-intensity of air transport compared to ship or train (e.g. Hines and Lang 1993). On the other hand, many economists argue that as trade improves the overall efficiency of economies, this also reduces various environmental impacts, and produces the wealth that can be spent on environmental improvements (e.g. Copeland and Taylor 2003).

A second element of the trade-environment connection concerns how trade is multilaterally managed. Whether or not trade itself necessarily increases environmental degradation, conflicts may arise between the imperative of reducing barriers to trade and that of reducing pollution. Economists frequently argue that environmental protection measures, which may include policy measures such as product standards, subsidies to particular products (renewable energy, for example), bans on certain production processes, and so on, in fact are usually designed as more conventional protectionist measures, designed to favour domestic industry over international competitors. More generally they worry that even if not protectionist in intent, the pursuit of environmental measures will reduce the volume of international trade on which general prosperity depends. This problem also illustrates the tension between a globalizing economy and an interstate system, as outlined above by realists and liberals. States are by now enmeshed in a set of globalizing pressures which constrain them (as well as enabling them to grow), but will often have powerful national pressures either to protect their own industries, or to protect particular ecosystems, species, etc.

There have been a number of iconic test cases in the context of the GATT and now the WTO. The most famous in the GATT is the Mexican 'dolphin-tuna' case. Since 1972 (in the Marine Mammals Protection Act), the USA had legislation protecting dolphins in the Pacific, which banned the use of 'purse-seine' nets by US fishing fleets, since they drown dolphins while catching tuna. In 1988, the USA used this legislation to ban imports of Mexican tuna, which still used this method. Mexico brought the case to a GATT Panel in February 1991. The Panel decided that the USA was not allowed to use domestic regulation to protect the environment outside its jurisdiction, not allowed to consider the methods of production as a consideration in decisions. This was widely seen as a precedent that makes a whole host of other regulations around the world contrary to GATT.

A more recent widely discussed case is the Shrimp Turtle case. In this case, India, Malaysia, Pakistan and Thailand brought a case to the WTO in 1998 to challenge a US law which banned imports of shrimp from those countries because they continued to use methods of shrimp catch which caught and killed sea turtles. However, in this case, taken by some as a sign of environmental progress in the WTO, the Panel, after requiring some changes in US legislation, found the US laws acceptable. Key here was the existence of multilateral agreements protecting sea animals, and ongoing negotiations between the USA and the countries concerned for further agreements. Nevertheless, while this is a sign of some progress, in particular concerning potential conflicts between Multilateral Environmental Agreements and the WTO, there are still remaining tensions between the management of trade and environmental concerns (Eckersley 2004: 35–39).

The second aspect of the global economy and its management often considered important is that of finance. While there is some attention to the power of private financial actors and flows in shaping environmental outcomes (e.g. Clapp and Dauvergne 2005: 214–216; Schmidheiny and Zorraquin 1996; French 1998) the majority of the debate here concerns the relationship between developing country debt and its management by the IMF and the World Bank. After the onset of the debt crisis in 1982, the power of those two institutions to direct economic policy in many developing countries increased considerably. They imposed a series of policy changes across the developing world, usually known under the rubric of 'structural adjustment' and later through the idea of the 'Washington consensus' (Williamson 1990). For many analysts, structural adjustment policies have produced a dynamic which profoundly aggravated environmental degradation.

There are a number of reasons for this. First, and perhaps most importantly, structural adjustment forced countries aggressively to increase exports while cutting imports. Given that to diversify economies away from the dependence on natural resources and agricultural products, on which most developing country economies had been based since the colonial era, would require importing capital goods and/or finance to build a manufacturing base, the requirement to slash imports meant that developing countries became ever more reliant on natural resources and foodstuffs for their export earnings. Structural adjustment thus produced an intensified attempt to increase exports of these products, thereby generating pressures which accelerated deforestation, soil erosion, over-use of marginal land, increased irrigation and water consumption, and so on. It also made developing countries more reliant on multinational firms for access to new seed strains and artificial fertilizers and pesticides in order to increase yields (intensified in the 1990s by new rules concerning intellectual property protection), with effects on water and soil quality as well as on the livelihoods of small farmers unable to pay for these inputs and thus compete in the new circumstances. The increase in deforestation in Brazil is particularly well documented (e.g. George 1992). The Brazilian government intensified its incentives to clear rain-forest land for cattle ranching in particular, building roads, turning a blind eye to violence against forest users, granting land to southern farmers and urban dwellers, and so on (Hecht and Cockburn 1990). Rates of deforestation in highly indebted countries have been systematically higher than those without great debt problems. More recently, the acceleration of deforestation in Indonesia can be associated with its economic problems after the 1997 crisis (Dauvergne 1998).

Much of these debates focus on whether an individual state can effectively pursue environmental policies in the context of these constraints from both the global economy itself and the way it is multilaterally managed. But many have also suggested that there may also be significant contradictions between the various multilateral environmental regimes which have been developed, and the economic regimes governing trade, investment, and finance. In particular, many have worried that the WTO rules have created a 'big chill' (Eckersley 2004), impeding the development of multilateral rules in various environmental contexts.

Many environmental regimes do explicitly function through trade restrictions. Most explicitly, the Convention on the International Trade in Endangered Species (CITES, 1975) operates not through restricting the killing of species but through restricting trading in them or in products derived from them. The Montréal Protocol on Substances That Deplete the Ozone Layer (1987) prohibits parties to the agreement from trading in ozone-depleting substances with non-parties. In the climate change context, commentators and participants have worried whether various types of policies proposed would contravene parties' obligations under the WTO. For example, were countries under the UNFCCC or Kyoto Protocol (or its successor) to agree to collectively impose a carbon tax, and then to impose a border adjustment tax on imports from countries not party to the agreement, those other parties could challenge this in the WTO context. It is worth noting, however, that despite these concerns, there have not yet been any challenges to trade-restricting clauses in multilateral environmental regimes. With CITES, this is relatively unsurprising since the convention effectively operates through the self-denial of exporting countries—an importing state (say Japan) is not in the position of refusing imports of elephant ivory from (say) Zimbabwe) in order to protect its own elephant industry. With the Montréal Protocol, however, it is reasonably clear that a good case could be made by non-parties that the trade-restricting clause was unfair under WTO rules, yet no state challenged the rule. This is not to say that challenges may not occur in the future, and it is certainly possible that a more subtle 'big chill' effect occurs—causing states to censor what types of policies they propose in order not to create tensions with WTO rules.

Many suggest therefore that dealing effectively with a range of environmental problems requires reform of global economic institutions. Proposals abound to reform institutions like the World Bank, to increase aid flows and direct them to environmentally sustainable forms of development, to reform trade rules to favour sustainable forms of production and consumption, to create a new global environmental organization to match the power of the World Bank and the WTO, and so on (e.g. Speth and Haas 2006; Biermann 2001). Others, more radically minded, argue for abolishing both global institutions and widespread global economic flows, and reorganizing economies at local levels (Hines 2000; Goldsmith and Mander 2001).

It is worth noting that at the heart of this set of debates is the first of the two conceptualizations of IPE outlined above. The fundamental problem is, in this view, that states pursue their own interests while the economy is globalizing. This latter globalization undermines attempts by states, individually or even collectively, to pursue environmental goals effectively. The lack of authority at the global level is one way of seeing this problem; alternatively it could be understood as an over-extension of the scale of the global economy, prompting suggestions to promote localization mentioned above.

STATES, FIRMS AND THE ENVIRONMENT

A second key question in the IPE of the environment concerns the relationships between states on the one hand and private firms on the other. The key problematic here is the power of business to shape and constrain the pursuit of sustainability. Many suggest that multinational companies (MNCs) are able to use their power—deriving not only from their size but also from their mobility, their ability to move investments around the world and thus play states off against each other—to evade or weaken environmental regulations, or that states avoid such regulation in anticipation of negative reactions from firms.

In contrast to the previous discussion, the concept of IPE here usually adopted is much more commonly the second, more critical one. Here, the fundamental interests of states and firms are regarded as largely coincident; while there may be specific conflicts, especially if social movements are able to get particular issues on the political agenda, in general states assume to be promoting business interests. Another way of seeing this contrast is to think through the buzzword of 'globalization'. In the previous discussion, this term is a synonym for 'global integration'—the world's markets becoming progressively more integrated with each other. The focus on MNCs and their power is more consistent with a notion of globalization as the global organization of corporate power, as reflected in such book titles as *When Corporations Rule the World* (Korten 1996) or *The Corporate Planet* (Karliner 1997).

At their heart, the claims about MNCs and the environment start from the proposition that the principal goal, even responsibility, of corporations, is to maximize returns for their investors. As such, they have no specific requirement to care about reducing resource use or pollution, and in fact specific incentives to offload the costs associated in particular with pollution onto other actors—people in other jurisdictions, society in general, future generations. At the same time, corporations have grown since the late 19th century into the principal form of economic organization in global capitalism. Consequently, their power to pursue their interests has grown.

MNCs now account in reasonably direct terms for large shares of global resource use and pollution. One central question is whether or not MNCs can 'go green'. Do the forces shaping MNC behaviour mean they inevitably engage in environmentally destructive behaviour, or are there countervailing dynamics that mean (at least in some circumstances) their environmental impacts can be significantly reduced? There is now a substantial debate on 'the greening of corporations' (e.g. Prakash 2001). For some, at the most optimistic end, the environmental impacts produced by firms are signals of their inefficiency—that if they act to reduce their consumption of raw materials, their emissions of pollutants, and so on, they will simultaneously reduce their costs (in money terms), producing a 'win-win' situation. Others are less optimistic, but nevertheless suggest that a number of features—the costs of resource inputs, the dynamics of technical change, pressure from consumers and investors, concerns about legal liability, for example—create possibilities of virtuous cycles for improving environmental performance. Key here is the role of regulation—the extent to which the stimulation of this behaviour requires regulatory change to shape incentives facing firms, which of course begs the question of corporate power to shape and resist regulation in the first place. Some (e.g. Hawken et al. 1999) suggest much of the change can occur (and is occurring) with only limited push from government, just firms pursuing the bottom line, but most others (e.g. Karlson, Hargroves and Smith 2005; Porter and van der Linde 1995) suggest regulatory change is in fact a crucial part of the picture.

A second question is the specific one concerning how MNCs shape the distribution of pollution. One particular debate in this regard is in the notion of 'pollution havens'. This debate concerns whether or not the activities of MNCs effect a migration of dirty industries from rich countries to poor countries, and therefore an imposition of the environmental costs of consumption in rich countries on the poor. There is little debate that some sort of 'migration of dirty industries' occurs, but more of a debate about whether this is stimulated by deliberate policy activity in some developing countries (the country deliberately reducing environmental regulation to act as a 'pollution haven') and whether or not the MNC activity involved should be regarded as exploitative or not (for varying views, see for example Clapp 1998b, 2002; Wheeler 2001; Neumayer 2001).

A third aspect of this debate concerns the power of MNCs and its political consequences. Their power, however, means that they can act to defend their interests either when they think they are threatened by environmental regulation, or to promote environmental regulation which advances their interests. The way that corporations have acted to resist environmental regulations, to do with pollution of air, water or soil, to change production processes to make them more 'eco-efficient', and so on, is well documented (e.g. Doyle 2000; Gonzalez 2001), as well as attempting to undermine environmentalism politically (Rowell 1996). A classic case concerns climate change, where a group called the Global Climate Coalition for a long time was the largest corporate group organizing on climate change, representing principally coal, oil, automobile, steel and electricity interests (and primarily in the USA), which was successful (at least until recently) in preventing governments from acting to reduce emissions substantially, principally through its activities in the USA (e.g. Newell 2000; Newell and Paterson 1998; Newell and Levy 2005).

It is not, however, simply a question of MNCs always resisting action on environmental problems. At times, they act to shape regulations in ways that pursue their interests. An interesting case is the Montréal Protocol, the reaching of which was made significantly easier by shifts in the strategy in particular of DuPont and to a lesser extent ICI, the world's two largest CFC producers. Around 1985, these firms realized that they may be able to increase their overall market share by backing action to limit CFC production, knowing that they had competitive advantage in the market for alternatives to CFCs (Oye and Maxwell 1994). Again, more recently, climate change governance has been increasingly shaped less by the resistance of fossil fuel producers and heavy consumers, and more by financial actors seeking to gain from the emergence of the 'carbon market' (Matthews and Paterson 2005).

The final point about MNCs and global environmental governance is that there are now a number of instances where MNCs are acting collectively to create such governance on their own (Falkner 2003). This 'privatisation of governance' (Clapp 1998a) works in different ways in different contexts, but broadly means firms, or international organizations dominated by firms, agreeing rules, norms, to guide their production practices and trading behaviour. A significant political dynamic here is to forestall (inter)state regulation (Newell 2001) but it can also be designed to enable legitimation in 'green markets', and at times, as with the CFCs case, to enhance market share and power. The case of the ISO14000 series of environmental management standards is widely mentioned here (see Clapp 1998a for the best analysis), through which firms which try to get certified have to show they have a series of environmental management systems in place.

ECONOMIC GROWTH AND SUSTAINABILITY

A third key question is that of economic growth. A classic environmental claim concerns the contradiction between a planet with finite resources, and finite ability to absorb anthropogenic pollution without ecological collapse, and an economic system premised on infinite expansion of production and consumption. The most famous example of this argument is in the book *The Limits to Growth*, published in 1972 (Meadows et al. 1972). This report, financed by the Club of Rome, a group of leading industrialists, used a computer model of the dynamics of population growth, economic growth, technological change, and so on, on the back of assumptions about available resources and ecosystem absorption capacities, to suggest that during the 21st century the world would experience what they called 'overshoot and collapse'—the exceeding of a variety of ecological limits leading to collapses in food production, industrial production, and human population.

There was a fierce debate in the 1970s and 1980s in particular about the notion of limits to growth. Many of the specific predictions of the Meadows et al. team turned out to be manifestly incorrect. For some (e.g. Simon 1981; Simon and Kahn 1984) this was used as evidence that economic growth could continue indefinitely. As some resources may become scarce, the prices would go up, prompting their replacement with other less scarce resources, while technologies would change to increase the efficiency of their use or to create closed-loop systems of recycling and reuse. For Simon in particular, the ultimate resource was human ingenuity, which he regarded as infinitely able to adapt to particular crises without threatening the overall expansive system. But for many others (e.g. Daly 1977; Henderson 1981) the fact that the Meadows et al. computer model predictions were incorrect does not necessarily lead to the conclusion that a finite planet and a growth-oriented system are compatible. There is still much evidence that the expansion of the global economy continues to consume an ever-greater portion of global resources, and consequently to breach a series of limits of ecological capacity—for example in the climate system, various forest ecosystems, or soil erosion and fertility (for an integrated assessment of the ways these limits are being breached, see Loh and Wackernagel 2004).

For IPE, at least its critical variant, the key point here is then to identify the driver behind global economic growth. Understanding the capitalist character of the contemporary global economy is crucial here. Fundamental to capitalist society in this context is the inherent (but unstable) dynamism produced by the form of exploitation of human labour that defines the system. Capitalism can be most fundamentally defined by the wage-labour relation—those engaging in direct production meet their subsistence needs by selling their labour power in return for a money wage. For the present purposes, this entails three things. First, it is highly efficient as a form of labour exploitation compared to previous forms, enabling employers (capitalists) to drive down wages to subsistence levels and thus force increases in the work performed and, thus, profits realized. Second, it created in-built incentives for technological innovation since employers saw labour as a cost to be minimized. Third, the growth patterns are unstable, operating through booms and slumps, and thus states have evolved with stabilizing growth as one of their key objectives and means of legitimation.

At the heart of the limits-to-growth question then, from an IPE point of view, is the challenge it lays down to the capitalist organization of the global economy. This clearly makes dealing with environmental questions highly problematic, since states, firms, many social movements and other actors take growth for granted as an imperative. States have coevolved with capitalism, and are structurally impelled to promote growth—intervening in times of crisis (e.g. recession) to return economies to a growth path, maintain the basic conditions of money (rules of private property and money, the disciplining of labour). As the economy has 'globalized', much of the activity designed to promote growth has similarly shifted to global levels, co-ordinated through the panoply of global economic institutions such as those mentioned.

Environmentalism has thus frequently appeared in economic discourse as a constraint on growth. Even despite the optimistic projections of Simon and others, a dominant reaction of many to claims by environmentalists, whether couched in the language of limits to growth or not, has been that they are 'unrealistic', or even 'immoral', as they threaten economic growth and prosperity. Many concrete environmental policy debates are dominated by a question of the impact of this or that regulation on rates of economic growth. At the international level, this dynamic has been particularly marked in a North-South context. At the first UN conference on the environment, in Stockholm in 1972, where the discourse was dominated by ideas like those in the *Limits to Growth* report, much of the debate centred around claims by representatives of developing countries that this amounted to a new form of colonialism, with industrialized countries attempting to 'pull up the ladder behind them', using environmental debates to block development in the South (see e.g. Thomas 1992 on this debate).

Much of the development of discourse about the environment-development relationship has been driven by these two themes. On the surface, it has been produced by the need in diplomatic and geopolitical circles to negotiate this North-South dynamic. During the 15 years after Stockholm, this morphs into what is still the dominant notion of 'sustainable development', popularized in the Brundtland report of 1987 (WCED 1987). Brundtland and later ideological developments since then (see Bernstein 2001 for details) were driven by a need to articulate the environmental *problématique* in a way which accommodated Southern demands for continued growth to lift their societies out of poverty. But on a deeper level, it was driven by a desire in all states (not just in the South), as well as in most transnational firms, and many in environmental movements, to try to suggest ways that growth and sustainability might be compatible. Given the capitalist imperative for growth, the limits-to-growth claim presented (and still presents) a fundamental challenge to the basic organization of the global economy. Whether the sustainable development political fudge is actually coherent is a separate question—there are many who regard it as fundamentally an oxymoron (Sachs 1999)—but politically it is the global political economy which has driven the development of this discourse.

CAN CAPITALISM BE GREENED? COMPETING PERSPECTIVES

As the previous section showed, the ultimate question in the IPE of the environment, if not environmental politics per se, is 'can capitalism go green?' Of course, there are many different perspectives here. Jennifer Clapp and Peter Dauvergne (2005) provide a very useful typology of them, suggesting that there are four principal ways that people orient themselves to this question. They identify 'market liberal', 'institutionalist', 'bioenvironmentalist' and 'social green' as the principal perspectives on this question.[1]

For market liberals, sometimes known as 'free market environmentalists' (Anderson and Leal 1991), the market is identified as a form of economic organization that is organized principally to maximize human freedom and prosperity. It is also fundamentally flexible, and can adapt to widely varying social and political conditions. Thus as societies begin to value the environment, this is reflected in the prices gained for environmental services and in the willingness to pay for environmental clean-up and the costs of avoiding pollution. They tend to argue that the origins of environmental degradation are less in 'market failure', but rather the failure to correctly allow markets to work effectively, most commonly by failing to allocate private property rights. They also tend to argue (e.g. Simon 1981; Beckerman 1974), that claims about the finite nature of the planet are overplayed, and that human ingenuity, especially if enabled by free market policies, will create technological innovation to overcome specific environmental problems. Market liberals thus argue that markets can themselves become effective and efficient means of pursuing environmental goals, either by shifting incentives in existing markets, through taxes on environmental bads, or (in the purer form of the argument) by the direct creation of markets for trading permits to emit various types of pollutant. In effect, this latter proposal consists of allocating private property rights (for example to emit carbon dioxide) and then to allow those rights to be traded in the market. Overall, therefore, market liberals argue there is no basic incompatibility between a global market economy and the requirements of sustainability.

Institutionalists share with market liberals a general belief in the compatibility of the market economy with the needs of the environment, but suggest that on its own the market does not have the capacity to steer societies in the right direction. Rather, the strengthening and adaptation of social and political institutions is necessary to guide markets towards sustainability. For institutionalists, markets fail to provide environmental goods in a number of ways. Markets never in fact work according to the ideal logic of economists: for example prices do not necessarily rise sufficiently to prevent exhaustion of a resource, nor always stimulate switching to alternative resources. As a consequence, the building of institutions to guide markets is necessary.

The institutionalist claim has become particularly prominent in relation to international environmental questions. In part this is due to the 'anarchic' character of international politics, and thus much of the debate in global environmental politics has been about the search for better international environmental institutions (Young 1994; Speth and Haas 2006). But the claim is also that even in an anarchic international system, there are many norms and rules which guide state behaviour, and also there is plenty of room for manoeuvre for the development of such norms. A basic environmental norm is that contained in Principle 21 of the declaration from the Stockholm Conference of 1972, which states that 'States have ... the sovereign right to exploit their own resources pursuant to their own environmental policies, and the responsibility to ensure that activities within their jurisdiction or control do not cause damage to the environment of other States or of areas beyond the limits of national jurisdiction.' A more recent development is that of 'common but differentiated responsibility', articulated in the Framework Convention on Climate Change, as well as elsewhere, which identifies the differential responsibilities of industrialized and developing countries in relation to responding to climate change. These general norms are then embedded in the specific contexts of the myriad international environmental regimes on a huge range of issues. The general institutionalist argument is that these institutions at least have the potential to shape the development of the global economy in the direction of sustainability.

At the other end of the spectrum, both bioenvironmentalists and social greens reject the idea that capitalism can be rendered sustainable. The former tend to place the emphasis on the need to adopt ecocentric or biocentric ethical systems which reject the project of 'dominating nature' in favour of some or other version of an ethical approach which places inherent, not just instrumental, value on nature. Others within this approach tend to argue by analogy about how 'natural' systems work, and that human economies and societies need to mimic these natural systems—for example keeping resources and effluents within watersheds (the bioregional approach), or by creating economies where all by-products become inputs to other processes (an approach also adapted within capitalist contexts as 'industrial ecology' or 'closed-loop production'). All variants of this approach are insistent that the limits of the Earth, both physical and ethical, require significant scaling down of economic activity.

Clapp and Dauvergne's final category of 'social greens' covers a range of ecological positions which all reject the possibility of a sustainable capitalism, but also argue that bioenvironmentalists tend to be anti-humanist, or at least underplay the social character of the environmental crisis—that it is the product of specific social systems (not just 'humanity' in general) and that it plays out in concrete contexts of huge economic inequality and power differentials. While there are many differences of nuance, especially concerning how to characterize the nature of the social system which generates the environmental crisis, this general argument is shared by many in European green movements, Southern ecological activists, ecosocialists and ecoanarchists, and many ecological feminists.

It is thus in this view the specific character of capitalism/patriarchy/imperialism/statism which is at the root of the environmental crisis. For those focusing on capitalism (most immediately relevant in the context of IPE), this is a system which requires growth for its reproduction, and thus is at the root of the cultural obsessions with growth at the macro-level, and ever-expanding consumption at the personal level. It is also a system which has expanded economic inequalities on both local and global scales enormously by comparison with previous economic systems. Finally, it is a system which confers increasingly concentrated power in the hands of an economic élite which is thus both able to ensure that its interests are protected politically and which can insulate itself from the ecological costs of its practices. People from this perspective tend to suggest that the 'corporate greening' discussed above is little but 'greenwash', since the underlying dynamics of the system within which corporations operate drive continuing ecological degradation. Some call this the 'treadmill of production' (Schnaiberg, Pellow and Weinberg 2002) that continues relentlessly expanding resource consumption and inhibiting tendencies to 'ecological modernization' (Mol and Spaargaren 2002).

CONCLUSIONS

Clearly, the question of the sustainability of capitalism is not going to be resolved in the immediate future. The debate turns on empirical questions of rates of possible technical change which might decouple growth from its environmental impacts, but also more fundamentally on conceptual questions such as the nature of the imperatives driving corporate and state behaviour or the capacity to generate political support for environmental measures in the face of corporate resistance, and on normative questions such as the ethical limits to the appropriation of 'nature' by human economic activity.

It is perhaps worth noting how Clapp and Dauvergne's four perspectives on environment and the global economy map onto the two broad perspectives on IPE outlined at the beginning. Their first two map neatly onto the liberal variant of the 'mainstream' approach to IPE outlined first. They both emphasize the normative value of market economies, and that in principle at least, such economies can successfully address the environmental crisis. Interestingly, the 'realist' variant is not really seen here. This version tends to be represented more commonly among some bioenvironmentalists, as in William Ophuls' (1977) environmental arguments for world government—his argument is that a sovereign states system cannot generate sufficient co-operation to deal with the environmental crisis. Bioenvironmentalists tend to reflect versions of both 'mainstream' and radical arguments about global political economy, occasionally referring to the problems of the power of corporations, or the overextended scale of global institutions as they argue for localization, but often rather tending to actually understand the global economy in fairly conventional terms—the interactions of markets, firms and states—and lacking the sense of capitalism as an integrated whole that the radical perspectives insist on. Social greens for the most part do have this sense, and clearly identify the 'problem' in terms of the dynamics of growth, the power of corporations, and so on.

This essay has tried to elaborate the central debates about environmental politics when one views it from the focus on the global political economy and the debates within the academic field of IPE. It has not of course tried to 'answer' the questions posed, but nevertheless to suggest why they may be among the most important questions to address in thinking politically about the environment and sustainability.

Bibliography

Anderson, T., and D. Leal. 1991. *Free Market Environmentalism*. Boulder, CO: Westview Press.

Beckerman, Wilfred. 1974. *In Defence of Economic Growth*. London: Cape.

Bernstein, Steven. 2001. *The Compromise of Liberal Environmentalism*. New York: Columbia University Press.

Biermann, Frank. 2001. The Emerging Debate on the Need for a World Environment Organization. *Global Environmental Politics*, 1 (1): 45–55.

Clapp, Jennifer. 1998a. The Privatization of Global Environmental Governance: ISO 14000 and the Developing World. *Global Governance*, 4 (3): 295–316.

Clapp, Jennifer. 1998b. Foreign Direct Investment in Hazardous Industries in Developing Countries: Rethinking the Debate. *Environmental Politics*, 7 (4): 92–113.

Clapp, Jennifer. 2002. What the Pollution Havens Debate Overlooks. *Global Environmental Politics*, 2 (2): 11–19.

Clapp, Jennifer, and Peter Dauvergne. 2005. *Paths to a Green World: the political economy of the global environment*. Cambridge MA: The MIT Press.

Copeland, Brian R., and M. Scott Taylor. 2003. *Trade and the Environment: theory and evidence*. Princeton: Princeton University Press.

Daly, Herman E. 1977. *Steady-state economics: the economics of biophysical equilibrium and moral growth*. San Francisco: W. H. Freeman.

Dauvergne, Peter. 1998. The Political Economy of Indonesia's 1997 Forest Fires. *Australian Journal of International Affairs*. 52 (1): 13–17.

Doyle, Jack. 2000. *Taken for a Ride: Detroit's Big Three and the Politics of Pollution*. New York: Four Walls Eight Windows.

Eckersley, Robyn. 2004. The Big Chill: The WTO and Multilateral Environmental Agreement. *Global Environmental Politics* 4 (2): 24–50.

Falkner, Robert. 2003. Private Environmental Governance and International Relations: Exploring the Links. *Global Environmental Politics*, 2 (3): 72–87.

French, Hilary. 1998. *Investing in the future: harnessing private capital flows for environmentally sustainable development*. Worldwatch Paper 139, Washington, DC: Worldwatch Institute.

George, Susan. 1992. *The Debt Boomerang*. London: Pluto.

Gill, Stephen, and David Law. 1988. *The Global Political Economy*. Brighton: Harvester Wheatsheaf.

Gilpin, Robert. 1987. *The Political Economy of International Relations*. Princeton, NJ: Princeton University Press.

Goldsmith, Edward, and Jerry Mander. 2001. *The Case Against the Global Economy: and for a Turn Towards Localization*. London: Earthscan.

Gonzalez, George A. 2001. *Corporate Power and the Environment: The Political Economy of U.S. Environmental Policy*. Lanham, MD: Rowman and Littlefield.

Hargroves, Karlson 'Charlie', and Michael H. Smith (Eds). 2005. *The Natural Advantage of Nations: Business Opportunities, Innovation and Governance in the 21st Century*. London: Earthscan.

Hawken, Paul, Amory Lovins and Hunter Lovins. 1999. *Natural Capitalism: The Next Industrial Revolution*. London: Earthscan.

Hecht, Susanna, and Alexander Cockburn. 1990. *The Fate of the Forest: Developers, Destroyers and Defenders of the Amazon*. New York: Penguin Books.

Helleiner, Eric. 1996. International Political Economy and the Greens. *New Political Economy*, 1 (1): 59–78.

Henderson, Hazel. 1981. *The Politics of the Solar Age: Alternatives to Economics*. Garden City, NY: Anchor Press/Doubleday.

Hines, Colin. 2000. *Localization: A Global Manifesto*. London: Earthscan.

Karliner, Joshua. 1997. *The Corporate Planet: Ecology and Politics in the Age of Globalization*. San Francisco: Sierra Club Books.

Keohane, Robert O. 1984. *After Hegemony: Cooperation and Discord in the World Political Economy*. Princeton, NJ: Princeton University Press.

Korten, David. 1995. *When Corporations Rule the World*. London: Earthscan.

Lang, Tim, and Colin Hines. 1993. *The New Protectionism: Protecting the Future Against Free Trade*. London: Earthscan.

Loh, Jonathan, and Mathis Wackernagel. 2004. *Living Planet Report 2004*. Gland: WorldWide Fund for Nature.

Matthews, K., and M. Paterson. 2005. Boom or Bust? The political-economic engine behind the drive for climate change policy. *Global Change, Peace and Security*, 17 (1): 59–75.

Meadows, Donella, Dennis Meadows, Jorgen Randers and William Behrens. 1972. *The Limits To Growth*. London: Pan.

Mies, M. 1986. *Patriarchy and Accumulation on a World Scale*. London: Zed Books.

Mol, Arthur P. J., and Gert Spaargen. 2002. Ecological Modernization and the Environmental State. *Research in Social Problems and Public Policy*, 10: 33–52.

Neumayer, Eric. 2001. *Greening Trade and Investment: Environmental Protection without Protectionism*. London: Earthscan.

Newell, Peter. 2000. *Climate for Change: Non-State Actors and the Global Politics of the Greenhouse*. Cambridge: Cambridge University Press.

2001. Environmental NGOs, TNCs and the Question of Governance, in *The International Political Economy of the Environment: Critical Perspectives*, edited by D. Stevis and V. Assetto, 85–107. Boulder: Lynne Rienner.

Newell, Peter, and David Levy (Eds). 2005. *The Business of Global Environmental Governance*. Cambridge MA: The MIT Press.

Newell, Peter, and Matthew Paterson. 1998. A Climate for Business: Global Warming, the State and the Capital. *Review of International Political Economy*, 5 (4).

Ophuls, William. 1977. *Ecology and the Politics of Scarcity*. San Francisco: W. H. Freeman and Co.

Oye, Kenneth, and James Maxwell. 1994. Self-Interest and Environmental Management. Journal of Theoretical Politics, 6 (4): 593–624.

Paterson, Matthew. 1999. Globalisation, Ecology, and Resistance. *New Political Economy*, 4 (1): 129–146.

Porter, Michael, and Claas van der Linde. 1995. Toward a New Conception of the Environment-Competitiveness Relationship. *Journal of Economic Perspectives*, 9 (4): 97–118.

Prakash, Aseem. 2001. *Greening the Firm: The Politics of Corporate Environmentalism*. Cambridge: Cambridge University Press.

Rowell, Andrew. 1996. *Green Backlash: Global Subversion of the Environment Movement*. London: Routledge.

Sachs, Wolfgang. 1999. Sustainable Development: On the Political Anatomy of an Oxymoron, in *Planet Dialectics: Explorations in Environment and Development,* 71–89. London: Zed Books

Schmidheiny, Stephan, and Federico Zorraquin. 1996. *Financing Change: The Financial Community, Eco-Efficiency and Sustainable Development*. Cambridge MA: The MIT Press.

Schnaiberg, Allan, David N. Pellow and Adam Weinberg. 2002. The Treadmill of Production and the Environmental State. *Research in Social Problems and Public Policy*, 10: 15–32.

Simon, Julian L. 1981. *The Ultimate Resource*. Princeton, NJ: Princeton University Press.

Simon, Julian L., and Herman Kahn (Eds). 1984. *The Resourceful Earth: a response to Global 2000*. Oxford: Blackwell.

Speth, James Gustave, and Peter M. Haas. 2006. *Global Environmental Governance*. Washington, DC: Island Press.

Strange, Susan. 1988. *States and Markets*. London: Pinter.

Thomas, Caroline. 1992. *The Environment in International Relations*. London: Royal Institute of International Affairs.

van der Pijl, Kees. 1998. *Transnational Classes and International Relations*. London: Routledge.

WCED 1987. *Our Common Future – Report of the World Commission on Environment and Development*. Oxford: Oxford University Press.

Wheeler, David. 2001 Racing to the Bottom? Foreign Investment and Air Pollution in Developing Countries. *Journal of Environment and Development*, 10 (3): 225–245.

Williamson, John. 1990. What Washington Means by Policy Reform, in *Latin American Adjustment: How Much Has Happened?*, edited by John Williamson. Washington, DC: Institute for International Economics.

Young, Oran R. 1994. *International Governance: Protecting the Environment in a Stateless Society*. Ithaca, NY: Cornell University Press.

Note

1. I follow their typology here, although I place slightly different emphasis than they do on the most important elements in each perspective.

International Organizations and the Global Environment

HANNES R. STEPHAN AND FARIBORZ ZELLI[1]

INTRODUCTION

The organizational network of global environmental governance (GEG) mirrors the complexity of the planet's manifold and overlapping ecosystems. Bursting onto the international stage in the 1970s, environmental issues began to be addressed by a series of new international organizations, most of them affiliated with the UN. Some of them, such as the UN Environment Programme (UNEP), were given a broad mandate, whereas others like the World Meteorological Organization (WMO) concentrated on a much more precise issue-area and have gained significant authority for their respective subfields. After the end of the Cold War, the rise of international environmental organizations has continued unabated. Yet the new institutions came to life in an already *institutionalized* context: some of the urgent tasks of management and co-ordination had already been allocated, and the newcomers often contributed to a growing trend towards organizational fragmentation.

For this essay, we have adopted a broad and inclusive definition of international organization that is none the less distinguished from two other types of international institutions, namely what Keohane (1989: 4) describes as institutions with explicit rules (international regimes) and institutions with implicit rules ('conventions'). In contrast, the organizations we study are bureaucratic actors and 'purposive entities' which are 'capable of monitoring activity and of reacting to it' and have been 'deliberately set up and designed by states' (ibid.: 3). They include not only fully-fledged 'organizations', but also UN commissions and programmes. Among the plethora of organizations with environment-related activities, we have restricted our analysis to those operating at the global level and have further selected those with either a clear environmental profile or a significant impact on global environmental governance.

In addition to our leitmotif of organizational fragmentation—which evokes the image of a mosaic of institutional elements—we have also taken account of current debates over mainstreaming and sectoralization. Thus, many of the organizations reviewed in this essay contain indications of the progress made towards a greater cross-sectoral integration of environmental concerns. For instance, the World Bank or the UN Development Programme (UNDP) now routinely address environmental factors in their decision-making, albeit with variable sincerity. Such insights feed into our concluding analysis of future trends and perspectives for reforming the system of global environmental organizations.

We begin our survey by describing a number of well-known global environmental conferences which provided the seedbed for the steady expansion of international environmental activities.

UNITED NATIONS-SPONSORED GLOBAL ENVIRONMENTAL CONFERENCES

International organizations neither emerge from nor exist in a political vacuum and they commonly rely on national governments' support for the negotiation and implementation of environmental agreements. However, the UN system, which accommodates the key globally operating environmental organizations, is clearly more than a simple 'tool' of its members; in particular, the institutions or fora it initiates at times count as significant actors in their own right. As Bennett and Oliver (2002: 25–26) have observed, '[i]f environmental sensibilities and regulation have developed, the discourse that has produced them has occurred within a diplomatic and legal framework set by UN commissions and conferences over the last thirty-odd years'. The progressive accumulation of norms, principles and action plans—commonly interpreted as elements of non-binding 'soft' law— began with the UN Conference on the Human Environment (UNCHE) at Stockholm in 1972. Despite concerns from the developing world about potential implications for economic development (Imber 1996), the summit mostly dealt with 'first-generation' environmental problems such as point-source pollution. Apart from the creation of UNEP, Stockholm produced a detailed action plan of environmental measures, a political declaration of 26 principles, and gave a genuine impetus to national policy-makers, often leading to the formation of national environmental ministries (Chasek 2000: 3).

By 1992, when Stockholm's successor—the UN Conference on Environment and Development (UNCED)—was convened, the bipolar world had just come to an end. Optimism and a thirst for action were palpable at the 'Earth Summit' in Rio de Janeiro and such feelings were not limited to the growing number of environmental NGOs that had come as observers. '[G]lobal environmental change,' ventures Vogler (2007: 435), had 'in some ways replaced fears of nuclear Armageddon.' On a substantive level, the forceful emphasis of the global South on development issues, already visible in Stockholm, had left a deep imprint on the international agenda. Seeking to integrate the demands of environmental protection and socio-economic progress, the concept of sustainable development—popularized by the Brundtland report in 1987—arguably embodied the 'central ideology of UNCED' (Imber 1996: 139) and it pervaded the 27 principles contained in the Rio Declaration. In addition, negotiators produced the *Agenda 21*, a 700-page non-binding action plan that has since continued to guide environmental policy-making at all governmental levels (Chasek 2000: 4). The Commission on Sustainable Development (CSD) was tasked with reviewing the progress towards its goals. Finally, UNCED also created treaties on climate change and on biodiversity which spawned several important protocols in the ensuing years—the most famous being the Kyoto Protocol. With hindsight,

UNCED's two major environmental conventions—the UN Framework Convention on Climate Change (UNFCCC) and the Convention on Biologocal Diversity (CBD)—marked a shift towards greater governmental control: the secretariats of these newcomers could not match the relative independence of some earlier global environmental conventions, such as the 1973 Convention on International Trade in Endangered Species (CITES).

When the World Summit on Sustainable Development (WSSD) at Johannesburg went underway in the summer of 2002, much of Rio's idealism had been exhausted. Developing countries were buoyed by the rising prominence of the UN's Millennium Development Goals (MDGs). The developed world, on the other hand, was affected by public apathy and increasing tension over both trade and environmental policies between the USA and the European Union (EU). Talk of a new mobilizing idea—a 'global deal' between North and South—was quickly abandoned in favour of finding practical ways of implementing previous, unachieved commitments through 'isolated delivery mechanisms' (Bigg 2003). More emphasis than before was placed on public-private 'type 2' partnerships with a view to supplementing official development aid (ODA), but this was not matched by a binding code for corporate responsibility. The topics of trade and poverty eradication arguably stole the limelight from the title theme of sustainable development and reduced environmental considerations to a restatement of existing agreements (von Frantzius 2004). A political declaration and the Johannesburg Plan of Implementation (JPI) contained over 30 targets on both development and environmental issues, yet specific instruments for implementation were largely absent (ibid.: 472).

Overall, given the limited achievements of the WSSD, there is an unmistakable impression of 'summit fatigue' among both political actors and academic commentators. If the grand 'show-biz' diplomacy of global summits is clearly flagging, it may be worth looking at the myriad ways in which the wider 'UN environmental machinery' (DeSombre 2006) has sought to tackle global environmental issues. One of the bodies set up to provide a more continuous organizational effort is the UN Commission on Sustainable Development.

COMMISSION ON SUSTAINABLE DEVELOPMENT (CSD)

A brainchild of UNCED, the CSD began its life with high expectations. Steeped in the 'spirit of Rio' and entrusted with the global pursuit of sustainable development, the new body's primary objective was to provide a follow-up to the summit's priorities and review the implementation of Agenda 21. The CSD is constituted by 53 member states which serve three-year terms. Its status as a functional commission of the Economic and Social Council (ECOSOC) does not endow it with specific powers or significant resources. Instead, it seeks to assist the sustainable development agenda by making recommendations to the UN system (by reporting to ECOSOC), monitoring national reports on implementation, and organizing multi-stakeholder dialogues with major interest groups and government representatives. This systematic inclusion of civil society organizations is illustrated by a long list of accredited observers (3,000 in November 2001), among them hundreds of NGOs (Wagner 2005: 105).

The crucial question to ask is whether the CSD has lived up to the hopes of its creators and whether its continued existence can be justified. Despite some patently useful work on freshwater and forests, influential ideas or a significant policy impact have not been among its achievements. Its recommendations often resemble restatements of decisions made in other international fora and the underlying assumption that 'if they talk about it, they will implement it' has proved to be unfounded (ibid.: 118). Furthermore, over the years, the CSD's agenda had become increasingly crowded and this frequently prevented a thorough discussion on the various propositions.

Serious engagement with civil society organizations is the major innovation consistently mentioned by commentators. Yet, in the absence of a clear policy focus there have been successive attempts at improving the Commission's performance. The 1997 reform of the CSD tried to streamline the agenda and remove some overlap with other UN fora, but only the improvements made after the 2002 WSSD review have brought about visible change. The CSD now observes a bi-annual negotiation cycle, with preparatory meetings in between remaining in a kind of 'exploratory' mode (ibid.: 112). While the leaner agenda and more modest ambition of 'facilitating' political and technical 'learning' have removed some overlap with other UN organizations (e.g. UNEP), they have not led to a 'rebirth' of the CSD as a major co-ordinator in the environmental field. The failure to reach agreement on climate change during the 15th CSD session in May 2007 has documented this ongoing lack of leverage and appeal. Overall, its recent depoliticization—achieved through a greater emphasis on expert meetings and problem-solving in collaboration with industry groups and NGOs—has made the CSD a constructive, yet unobtrusive addition to the UN's environmental machinery. The modest ambition behind the revamped CSD stands in marked contrast with the hopes invested in a strengthening of UNEP.

UNITED NATIONS ENVIRONMENT PROGRAMME (UNEP)

History

UNEP is rightly seen as the core of environmental activities within the UN system. Its 35-year-long history has been marked by a series of crises, notable achievements, and reorientations. A product of the 1972 Stockholm Conference and the UN General Assembly's Resolution 2997, the new agency (first headed by Maurice Strong) was essentially modelled on the United Nations Development Programme (UNDP). This arrangement was intended to match the 'interdisciplinary and complex nature of environmental problems' and accommodate a 'vast catalogue of recommended actions' (Thomas 2004: 57–59). From the beginning, governments recognized the cross-cutting nature of environmental problems. UNEP's status as programme rather than fully-fledged agency should thus not be read as a sign of early disregard or subordination (Ivanova 2005: 32).

This form of organization, however, also meant that UNEP's resources were being spread thinly across a whole range of issues. The programme's role as the UN's environmental conscience—both co-ordinating and catalysing global environmental activities—resulted in a broad selection of seven priority areas: human settlements and habitats (later turned into UN Habitat), human and environmental health, terrestrial ecosystems, environment and development, oceans, energy, and natural disasters (Downie and Levy 2000: 356). Following 'muscular' but contested attempts at system-wide environmental co-ordination in the 1970s, UNEP came into its own in the early 1980s when it instituted the 'Programme for the Development and Periodic Review of Environmental Law', also known as the 'Montevideo Programme'. This decision cemented UNEP's role in catalysing and developing international environmental law.

During the same period, UNEP was nevertheless becoming marginalized in other areas of environmental policy-making. The formation of the World Commission on Environment and Development (WCED) appropriated much of its normative power and unique environmental mandate (Conca 1995). In response, UNEP produced the report 'Environmental Perspectives to the Year 2000 and Beyond' which proved to be an important influence on the 1992 Earth Summit in Rio de Janeiro (WRI 2003: 143). The conference itself, however, was not an unmitigated blessing for UNEP. It certainly ushered in a series of budget increases, but it also broadened the agency's remit once again and agreed on the formation of the CSD. The availability of such alternative fora exacerbated the ensuing crisis in the mid-1990s (Wagner 2005). Elizabeth Dowdeswell, who replaced the long-time executive director Mostafa Tolba in 1992, presided over a period in which many countries were losing confidence in UNEP's capability. Its ambitious System-Wide Medium-Term Environment Programme (SWMTEP 1990–95) was regarded as a 'meaningless checklist' or a device for 'turf-grabbing' by many UN officials (Thomas 2004: 88). In 1997, matters came to a head when the USA, Britain, and Spain linked continued financial support to significant organizational reform (Karns and Mingst 2004: 476). The 1997 Nairobi Declaration broke the spell of decline and gave UNEP, headed by Klaus Töpfer from 1998–2006, a new lease of life. The new dynamism was further strengthened by the 2000 Malmö Declaration of environmental ministers—meeting as the Global Ministerial Environmental Forum (GMEF) for the first time—who sent out a strong message of concern in the run-up to the 2002 WSSD.

In the new millennium, UNEP has been a highly visible component of GEG once again, not least due to a stabilizing budget and the debate over an upgrading of its status. Its fifth executive director, Achim Steiner, who took office in June 2006, lauded his predecessor Töpfer for helping to 'stabilize the organisation and expand its operations' and pledged to continue on the basis of this legacy (UNEP 2007: 3).

Structure and Activities
The Governing Council (GC) reviews UNEP's progress and establishes its specific priorities. As UNEP is a programme under the aegis of ECOSOC, the

GC is expected to report to it directly. The GC's 58 members are elected for four-year terms by the UN General Assembly according to a regional formula.[2] The members meet annually and their decisions are best described as the driving force of UNEP's overall legal and operational activities. This does not, however, always translate into a clear or consistent framework of priorities because member states often insist on their own preferred projects (Ivanova 2005: 22). During the remainder of the annual cycle, the Committee of Permanent Representatives (CPR), located in Nairobi, provides political guidance and monitoring. The organizational trinity is completed by the GMEF which has been convened since 2000 and is tasked with giving UNEP a stronger, long-term programmatic direction and political leadership.

The day-to-day running of the organization is the business of the UNEP Secretariat in Kenya's capital Nairobi and six regional offices around the world. UNEP has a comparatively small professional staff of just over 900 employees. Its annual budget is dwarfed by sister agencies like UNDP and reaches about US $260m.[3] The annual core, 'non-earmarked' funding (known as the Environmental Fund) has been hovering at just below $60m. during the past few years. Not having the capacity or funds of a genuine delivering agency, UNEP needs partnerships with NGOs and other international organizations if it wants to go beyond catalysing and administrating international environmental law (Conca 1995). For instance, co-operative projects have been conducted with the World Meteorological Organization (WMO) on the atmosphere or with the Food and Agriculture Organization (FAO) and the World Health Organization (WHO) on freshwater quality (Karns and Mingst 2004).

Overall, as Thomas (2004: 18) suggests, UNEP's activities can be summarized by the 'four Cs' of compiling, convincing, catalysing, and co-ordinating. The task of compilation is related to the agency's original mission of representing a clearing house for environmental data and research at a time when such efforts were still in their infancy. Under a programme named 'Earthwatch', UNEP began to co-ordinate observation techniques and data analysis among all UN agencies as early as 1973. Through various dissemination mechanisms, such as the Global Environmental Information Exchange Network (INFOTERRA), the Global Environmental Monitoring System (GEMS), and its annual flagship publication 'Global Environmental Outlook' (GEO), it has sought to maximize the impact of its scientific analyses. The second brief—convincing the world to take action—is closely linked to the vexed question of effectiveness, which is discussed below. Besides its scientific authority and management expertise, UNEP's 'most basic skill' is diplomacy (Thomas 2004: 31). As a vital ingredient in the catalytic role which has seen the agency assume the mantle of a leader or broker in particular negotiations, it determines the success or failure of attempts at mainstreaming new concepts. In this respect, it is worth recalling that UNEP had taken up the notion of 'sustainable development' in the early 1980s—even before it was popularized by the WCED in 1987.

Finally, UNEP's co-ordination mandate is generally seen as a disappointment. A succession of interagency bodies have been entrusted with the objective of achieving more system-wide programmatic coherence and with mainstreaming environmental goals. The latest incarnation, the Environment Management Group (EMG), has the unambitious task of identifying synergies in the UN system and commands only a minuscule resource base (Ivanova 2005: 29).

Evaluation

The ineffectiveness of UNEP's 'Sisyphean' co-ordination mandate (Imber 1996) appears to imply a negative judgement on its general performance. In what some writers consider a 'feudal' UN system with a weak centre and strong 'baronial' independent agencies (ibid.: 150), UNEP is continually emasculated and does not even have nominal authority over the environmental conventions it has helped to set up. Fruitful collaboration is surely a regular occurrence, but the idea of co-ordination assumes direct guidance from a lead agency. Frequently, however, UNEP represents a mere adjunct to existing projects: for instance, in providing organizational functions to particular multilateral environmental agreements (MEAs) or by acting as a scientific adviser to the dominant partners (World Bank, UNDP) in the Global Environment Facility (GEF). Its subdued status has been weakened further by the creation of alternative fora, such as the CSD in 1992, and its politically desirable, but impractical location in Kenya, far away from the 'corridor politics' of UN hot spots in New York or Geneva.

Of course, there are also more promising findings about UNEP's performance, in particular with regard to its catalytic functions. The agency has excelled in its roles as 'agreement facilitator', 'negotiation manager', and 'regime administrator' (Downie 1995: 176). In line with the standard functions of international organizations, it has scheduled meetings at a propitious time and generated negotiation procedures that have helped the quest for compromise solutions. Moreover, it has occasionally entered the debate as a capable actor itself: during the negotiations on ozone depletion UNEP's executive director Mostafa Tolba abandoned the appearance of impartiality and began to refer to 'UNEP's interests'. He judiciously used UNEP's organizational powers and scientific knowledge to push for an adequate international agreement (D'Anieri 1995: 165–166). Finally, the case of the Regional Seas Programmes illustrates the possibility of overall leadership responsibility. Six of the 13 Regional Seas projects are directly administered by UNEP and have been reliant on its diplomatic skill and scientific argumentation as well as on a constant stream of funding (DeSombre 2006).

Yet a thorough assessment of the programmes' results yields a picture that is symptomatic of UNEP's general record over the last decades: marked environmental improvement is difficult to ascertain, even if the measures agreed have surely helped to slow the pace of deterioration. '[U]seful but not dramatic work' (DeSombre 2006: 19) may well be a fitting description of both the Regional Seas Programmes and UNEP's impact on the wider area of global environmental governance. More joint planning and activities with its sister agency UNDP would arguably enhance the financial and political clout of the environmental sector in international politics.

UNITED NATIONS DEVELOPMENT PROGRAMME (UNDP)

Although the United Nations Development Programme is neither by mandate nor in practice predominantly geared towards environmental protection, its financial importance for tackling associated issues at the project level is undeniable. In 2005, 11% of the programme's US $3,000m. portfolio was spent on projects under the label of 'Energy and Environment', equalling $326m. (UNDP 2006: 4)—which is more than UNEP's total funding for the same period. The UNDP dates back to a resolution adopted by the UN General Assembly in November 1965. It was given the mandate to assist capacity-building in developing countries with a view to pursuing key objectives, which today include poverty eradication, democratic governance, crisis prevention and recovery as well as combating HIV/ AIDS. With 3,300 staff located at headquarters in New York or in one of the programme's 135 country offices, and with field activities in 166 countries, UNDP is the largest existing multilateral organization for technical assistance and co-operation (Biermann and Bauer 2004: 7).

UNDP's environmental role is largely defined by its function as an implementing agency of associated global funding mechanisms, namely the issue-specific Montréal Protocol's Multilateral Fund or the cross-cutting Global Environment Facility (GEF). Receiving 30% of the former's funding, UNDP has supported the phasing out of ozone-depleting substances in developing countries through technical assistance, direct investment, feasibility studies and demonstration projects (DeSombre 2006: 115). UNDP's role in the GEF has so far seen the management of 1,750 projects in more than 155 developing countries. To finance such projects, in 2005 alone, UNDP secured US $284.5m. from the GEF, but also attracted $1,020m. in co-financing from governments and donors (UNDP 2006: 16). The environmental reputation of the programme also rests upon renowned initiatives, for instance the Capacity 21 programme for storing and disseminating ecological data in developing countries, or the $1.7m. MDG carbon initiative, launched in February 2007, to install a pilot carbon trading scheme in China.

Due to these diverse and widespread field activities, UNDP has rightfully been praised as 'a pragmatic complement to UNEP's global environmental treaty-making efforts', thereby promoting the idea of 'mainstreaming', i.e. the integration of environmental concerns into its development agenda (WRI 2003: 144). However, this assessment needs to be balanced by considering ongoing inter-agency tensions and turf wars within the UN environmental machinery. Despite common projects with UNEP (e.g. the 2007 launch of the Poverty and Environment Facility to support the implementation of the Kyoto Protocol in five African countries), observers have pointed to a historically grown lack of co-ordination which 'pre-dates the integrative concept of sustainable development' (Biermann and Bauer 2004: 19). Given continuous internal reforms, a significant rise of non-core resources, and several shifts in environmental priorities over the last decade (ibid.: 9, 20), it remains to be seen to what extent the programme can both follow its mainstreaming approach and achieve a better division of labour with UNEP.

WORLD BANK

A quite different approach to multilateral development assistance has been adopted by the World Bank. Unlike UNDP's grants-based assistance, the World Bank—as well as four regional multilateral development banks—supports projects with loans to be repaid. Moreover, the World Bank features a lower level of inclusion of developing countries and non-governmental organizations with regard to decision-making or disclosure of information. Since its establishment in 1944 under the name of 'International Bank for Reconstruction and Development' (IBRD), the Bank has steadily expanded and today comprises a closely associated group of five development institutions with up to 185 members. Its mission has gradually evolved from post-war reconstruction in the early days to world-wide poverty alleviation—hence also touching upon environmental issues. In 2005, the Bank spent US $2,490m. on environmental and natural resource management, equalling 11% of its overall portfolio.[4]

Apart from this extensive lending for environmental projects, some of the Bank's success stories are owed to its capacity for 'convening governments and setting guidelines' (WRI 2003: 143). Examples for this effective 'soft law' approach range from the initiation of a dialogue among logging industry leaders on sustainable forestry in 1998 to the 'Equator Principles', i.e. environmental investment-guiding criteria based on World Bank standards. Another prominent case is the launch of a multi-stakeholder dialogue which led to the 1998 creation of the World Commission on Dams. The commission released principles and guidelines on future water and energy decision-making; notably, however, the principles were later rejected by the Bank's board of directors (ibid.: 170; Dingwerth 2005).

On the other hand, the Bank has attracted strong criticism for its contradictive agenda, mainly because some of its conventional projects (for instance the promotion of the use of fossil fuels) can severely undermine the positive results achieved with GEF funding (DeSombre 2006: 160). Critics have also pointed out that the Bank's lending policy is biased towards economic profit, which creates difficulties for many environmental projects on problems resulting from unpriced externalities (ibid.: 157). Moreover, the practice of drafting so-called Poverty Reduction Strategy Papers (PRSP) for recipient countries has been interpreted by UNDP as a redressed version of the Bank's highly controversial structural adjustment conditionalities (Biermann and Bauer 2004: 11).

GLOBAL ENVIRONMENT FACILITY

Mistrust towards a potentially hidden conditionality has also accompanied another multilateral environmental financing institution: the Global Environment Facility (GEF). Having been inspired by discussions in the World Bank, the facility's pilot phase between 1991 and 1994 saw continuous tensions between developing countries and the USA over structural reforms (DeSombre 2006: 157). The result of these debates was a new type of international institution, 'an amalgamation of traditional features of UN and Bretton Woods institutions' (Streck 2002: 130ff.). As an open-ended funding mechanism for global environmental issues, the facility is more transparent and democratic than the World Bank thanks to a double voting system, independent reviews and a significant participation of over 700 NGOs (WRI 2003: 153). Further distance to the World Bank was assured by designing GEF as a provider of grants—instead of loans—and by naming UNEP and UNDP as additional implementing agencies.

The facility is mandated to finance incremental costs, i.e. new and additional funding which would not have been provided by other sources. This guideline has been criticized as failing to address 'the underlying political causes of environmental degradation in developing countries' (DeSombre 2006: 160). This notwithstanding, between 1991 and 2004, the GEF allocated an impressive total of US $6,800m. in grants, and could also leverage another $24,000m. in co-funding by governments, international organizations and private entities. With these resources, the facility has supported over 1,900 projects in more than 160 developing countries and countries with economies in transition.[5] More than half of this money was invested in the domains of the two Rio conventions—biodiversity loss and climate change—followed by four other GEF focal areas: international waters, ozone depletion, land degradation and persistent organic pollutants.

The GEF deserves special credit for this allocation record. Furthermore, after its early restructuring, the facility has undoubtedly become one of the most adaptive and transparent international institutions and displays a relatively high degree of North-South co-operation. Despite these achievements, it still has a difficult standing among some of its 177 members. Resistance to its work originates from both camps: whereas some of its sponsors have repeatedly failed to meet their funding obligations, some of the recipients resist the increasing scope of the facility's activities and are unwilling to distribute funds among too many focal areas (DeSombre 2006: ibid.). In light of this opposition, some critics have voiced doubts about the facility's innovative impulses. The GEF has to make considerable co-ordinative efforts in order to preserve a reasonably peaceful working relationship between implementing agencies and associated organizations—a role which does not grant much leeway for supporting experimental or cutting-edge projects. On a final cautionary note, whether or not specific projects can count as successes, the GEF's role has sometimes been criticized for 'greenwashing' the impact of the World Bank's ongoing investment practices (Young 2002).

WORLD TRADE ORGANIZATION (WTO)

History

As compared to the above institutions, the WTO is different in several regards: first of all, as a 'related organization' it is independent and hence far more detached from the UN system than programmes such as UNEP or UNDP and specialized agencies like the World Bank[6] or the GEF, which is administered by the former three. Moreover, the WTO has no proactive environmental mandate, neither for financial nor technical assistance. Its environmental role is exerted in an *ex post* or indirect manner, which none the less has significant impact due to the organization's considerable enforcement capacities.

This does not imply that environmental issues have not materialized in the organization's structure or documents. In fact, sustainable development is recognized as a key objective in the preamble of the WTO agreement. And institutional arrangements date back to pre-WTO times: in November 1971, on the verge of the UNCHE conference, the General Agreement on Tariffs and Trade (GATT) established the Group on Environmental Measures and International Trade (EMIT) in order to account for the trade implications of environmental policies. None the less, due to a lack of requests from the contracting parties, EMIT never convened in the first 20 years after its establishment. It was thus only the late 1980s which saw a second environmental debate take place within the architecture of the GATT. This second debate 'came at an awkward time for GATT signatories, since the Uruguay Round entered a deep crisis in the early 1990s and the agricultural dispute between the USA and the EU threatened to scupper the talks' (Santarius et al. 2004: 10). Though advocated by major industrialized countries, any comprehensive approach to ecological standards was blocked by developing countries which interpreted them as a disguise for protectionist measures (Eglin 1998: 252).

Structure and Activities

The major institutional manifestation of the WTO's environmental role is the Committee on Trade and Environment (CTE). The committee has a standing agenda and includes all current 150 WTO members as well as several observers from intergovernmental organizations (but not from NGOs) who gather at least two times a year for formal meetings plus further informal ones if needed. Its chief mandate is to ensure a positive interaction between trade and environment measures inside and outside WTO law—and to recommend appropriate modifications to the latter where necessary. The CTE is supported by the WTO Secretariat's Trade and Environment Division, which provides technical assistance to WTO members, reports to them about discussions in other intergovernmental organizations and maintains contact with non-governmental actors.

Despite these bodies and their mandates, it is not accurate to speak of a proper WTO environmental *policy*. The Trade and Environment Division is merely performing a service function while the WTO Secretariat has not been endowed with any competency to set its own environmental agenda (Bernauer 1999: 132ff.). Similarly, the CTE is anything but proactive on ecological matters: first of all, the committee's mandate is *not* to tackle free trade's impact on the environment; instead, it is supposed to act under exactly reversed premises and address the effects of environmental measures on trade policy (Santarius et al. 2004: 48). Second, the CTE does not consist of independent agents but of governmental representatives and its reports rest upon consensual decision-making. This lack of environmental momentum from within the WTO was desired by its creators, bearing justice to concerns voiced by developing countries who feared a 'green' conditionality for market access.

Given these intended shortcomings, the environmental agenda of the WTO is mostly shaped through a different channel: via the conflict of WTO law with domestic and international environmental regulations, and via the respective judicial interpretation and settlement of these conflicts (cf. Zelli 2006). In terms of quantity, the WTO Dispute Settlement Body (DSB) has constantly broadened its ecological agenda over the years—through decisions on topics ranging from species protection via air pollution to consumer and health standards. This development mirrors the general extension of jurisdictional scope during the transition from GATT to WTO: today, no fewer than 60 legal instruments under the auspices of the WTO cover a multitude of different policy fields, from agriculture to labour rights or from international finance to telecommunications (cf. Sampson 2005: 128ff.). The DSB and its two-layered system—consisting of the Panel and the Appellate Body (AB)—cannot issue reports on their own initiative, but the member states can invoke the DSB in order to block the implementation of other countries' ecological policies. Hence, the DSB substantially differs from the WTO's political bodies because it is not caught in a stalemate among countries and can reach final decisions through independent procedures.

In terms of substance and quality, one can observe an increasing tendency towards more flexible and integrative decisions. This concerns two key types of contested environmental standards: on the one hand, the precautionary principle, for instance addressed in a famous case on beef treated with hormone growth promoters (1998); and, on the other hand, provisions related to production methods, for example the US import bans based on fishing methods. Such non-trade preoccupations have gradually become integrated into the decisions—either through demands for multilateral negotiations and agreements in order to specify WTO law (as in the 1998 *US – Shrimp* report)[7] or through the recognition of the actual objectives of trade-restrictive measures (especially health issues, as in the *EC – Asbestos* decision). However, given increasing protests by WTO members about the AB's flexible interpretation of the agreement (Sampson 2002: 23), only time will tell whether this tendency towards more environmentally sound rulings will prevail.

Apart from these conflicts over domestic environmental regulations, a number of noteworthy overlaps exist between WTO law and MEAs. Some of the trade-related measures of the Montréal Protocol on ozone depletion, for instance, collide with the WTO principle of most favoured nation treatment 'by banning the import of various substances on the basis of the status of the country of origin' (Werksman 2001: 183). Moreover, the Kyoto Protocol might get into conflict with WTO law on a number of aspects, one of them being its constraints on the trade in carbon emissions (Chambers 2001: 103).

Evaluation

It is crucial to deny the merely theoretical character of WTO–MEA conflicts on two grounds: first, the current lack of legal disputes may be due to the fact that the majority of the MEAs in question have only been adopted within the last 15 years, and some of them have either not yet or only recently entered into force. Second, although there are no judicial controversies, the shadow of WTO law and its strong dispute settlement system may well provoke anticipatory conflicts or 'chilling effects' (Stillwell and Tuerk 1999; Eckersley 2004), whereby MEA negotiators refrain from specifying more ambitious trade-relevant measures or face a country's refusal to ratify an agreement or protocol (Pauwelyn 2003: 237ff.).

At present, any solution or regulation of these conflicts and overlaps between WTO law and domestic or international environmental rules seems improbable. There have been several initiatives, including a 1999 co-operation agreement among WTO and UNEP secretariats which launched a regular exchange of information on legal issues. Moreover, a 'trade and environment' section has been included in the WTO's 2001 Doha Declaration (Articles 31–33). Article 32 extended the CTE's mandate towards 'the effect of environmental measures on market access', the environmentally relevant provisions of the TRIPS Agreement (which, for instance, overlap with CBD rules on access to genetic resources and benefit-sharing) and 'labelling requirements for environmental purposes'. Pursuant to this explicit request for compatibility, a CTE Special Session (CTESS) was to discuss a number of models for harmonizing WTO law and the trade-related measures of MEAs. However, mirroring the overall crisis of the Doha Round, the first CTESS as well as its follow-up meetings—e.g. on the liberalization of environmental goods and services—have stimulated little agreement among WTO members on the further co-ordinative process.

OTHER INTERNATIONAL ORGANIZATIONS IN GLOBAL ENVIRONMENTAL GOVERNANCE

The previous sections have introduced major, globally operating international organizations and bodies which either have an environmental mandate or have otherwise exerted significant influence on environmental issues and policies. With the exception of the WTO, all of these organizations represent core components or affiliated institutions of the United Nations system. In addition, Table 1 lists a number of further international organizations engaged in environment-related activities.[8] Not surprisingly, these organizations are also linked to the UN. All but the last two have the status of a specialized agency (i.e. of an autonomous organization working with the UN).

Table 1: Selected International Organizations with Environmental Activities

Organization	Est.	Function	Website
Food and Agriculture Organization of the United Nations (FAO)	1945	FAO is the lead UN agency responsible for assessing the state of global agriculture, forests, fisheries, and for promoting sustainable development and harvest of these resources.	www.fao.org
United Nations Educational, Scientific and Cultural Organization (UNESCO)	1945	UNESCO promotes collaboration among nations through education, science, culture, and communication in order to further universal respect for justice, for the rule of law, and for human rights.	www.unesco.org
United Nations Industrial Development Organization (UNIDO)	1966	UNIDO works to strengthen industrial capacities of developing and transition nations with an emphasis on promoting cleaner and sustainable industrial processes.	www.unido.org
International Atomic Energy Agency (IAEA)	1957	The IAEA serves as an intergovernmental forum for scientific and technical co-operation in the peaceful use of nuclear technology, promoting nuclear safety and non-proliferation.	www.iaea.org
International Maritime Organization (IMO)	1948	The IMO is responsible for improving maritime safety and preventing pollution from ships.	www.imo.org
World Health Organization (WHO)	1948	The WHO catalyses international co-operation for improved health conditions, including a healthy environment.	www.who.int
World Meteorological Organization (WMO)	1950	The WMO co-ordinates scientific efforts in global weather forecasting and conducts research on air pollution, climate change, ozone depletion, and tropical storms.	www.wmo.ch
United Nations Population Fund (UNFPA)	1969	The UNFPA assists countries in providing reproductive health and family planning services, formulates population strategies, and advocates for issues related to population, reproductive health, and the empowerment of women.	www.unfpa.org
Intergovernmental Panel on Climate Change (IPCC)	1988	The IPCC was established under the auspices of UNEP and the WMO to assess scientific, technical, and socio-economic information relevant for the understanding of climate change, its potential impacts, and options for adaptation and mitigation.	www.ipcc.ch

CONCLUSIONS

The international bodies and agencies which have been portrayed in this essay differ with regard to several dimensions, including the breadth of their mandate (environmental protection, sustainable development, or non-environmental issues) as well as their agenda and predominant policy approach (funding, technical assistance, rule setting or rule enforcement, etc.). Another distinctive criterion is the position of these organizations with respect to the UN; it is intriguing that—with the exception of the WTO—all of them are somehow linked to the UN system, albeit in different roles. Hence the observed variety of organizations in global environmental governance is mostly rooted in the complexity of the UN environmental machinery which, in turn, 'reflects the complexity and diversity of environmental issues themselves' (WRI 2003: 141). This observation notwithstanding, one should not judge this decentralized arrangement as an inevitable necessity, let alone welcome it as an overtly harmonious 'symphony' of organizations (ibid.: 139). For sure, the variety of platforms has produced numerous benefits, among them: raising awareness and generating information on a range of environmental problems and policies, mobilizing expertise from scientists and NGOs, providing international negotiating fora, making significant contributions to international environmental law, and building capacities to implement environmental policies in the developing world (ibid.: 141ff.).

Yet, on the other hand, the institutional fragmentation implies overlapping mandates and, more importantly, it entails considerable shortcomings in co-ordination: more often than not, the various institutions have restricted co-operation to a well-defined number of issues, and interagency 'turf battles' over competencies and resources are a constant occurrence. Most prominently, 'other UN bodies have refused to accept UNEP's mandate to co-ordinate all environmental activities in the UN system due to 'institutional seniority'. A number of UN agencies [...] possessed environmental responsibilities before UNEP was created and thus feel less of a need to defer to UNEP' (Ivanova 2005: 25). Apart from the high transaction costs arising from such institutional incoherence, this patchwork is not capable of playing the role of a strong advocate for global environmental concerns *vis-à-vis* governments or non-environmental organizations. As a result, the various bodies of the UN environmental machinery have to compete for scarce contributions from national governments, while failing to convince other organizations to open their portfolios more extensively for environmental concerns.

Thus, the two ongoing debates on global environmental governance we mentioned at the outset of this chapter—fragmentation vs. centralization and sectoralization vs. mainstreaming—are clearly inter-related. Merging both discussions has inspired calls for a centralized and cross-cutting World Sustainable Development Organization, or, with less mainstreaming zeal, for a UN Environment Organization—a centralized, but issue-specific authority (Biermann and Bauer 2006). This essay has implicitly made a similar case, by sketching the strong impact of the world trade regime on a largely toothless mosaic of environmental institutions and regulations. However, the section on the WTO has also revealed that creating a centralized counterweight is no reliable panacea. The WTO itself has repeatedly been dogged by conflict among its member states, quite similar to the stalemates which keep undermining the co-ordination among multilateral environmental organizations. The real difference is the WTO's strong dispute settlement mechanism which can temporarily circumvent such standstill and exerts an unprecedented influence on domestic and international policies. Thus, in order to play an effective role in 'Earth system governance' (Biermann 2007), a future world environment organization would need to be endowed with comparable dispute settlement and enforcement capacities.

Meanwhile, on a less ambitious but more realistic scale, international environmental organizations should try to maximize the synergistic potentials of their overlapping tasks. They could do so through enhanced mainstreaming and division of labour at the project level, and through bolder co-operation agreements at the organizational level. In addition, striving for cross-issue package deals among country coalitions might break negotiation impasses within and between organizations: governments could more actively link environmental issues with non-environmental concerns – especially with issues of 'high politics' such as security or trade. Undoubtedly, such integrative or mainstreaming attempts will have to walk a thin tightrope: improving interorganizational co-ordination while making sure that the environmental component is not diluted or absorbed by other concerns.

Bibliography

Bennett, A. L., and J. K. Oliver. 2002. *International Organizations: Principles and Issues*. Upper Saddle River: Prentice Hall.

Bernauer, T. 1999. Handelsliberalisierung und Umweltschutzpolitik: Konflikte und Synergien. In *Handel und Umwelt: zur Frage der Kompatibilität internationaler Regime*, edited by T. Bernauer and D. Ruloff, 118–140. Opladen, Germany: Westdeutscher Verlag.

Biermann, F. 2007. 'Earth System Governance' as a Crosscutting Theme of Global Change Research. *Global Environmental Change*, 17 (3–4): 326–337.

Biermann, F., and S. Bauer. 2004. United Nations Development Programme (UNDP) and United Nations Environment Programme (UNEP). Berlin, Germany: WBGU. http://www.wbgu.de/wbgu_jg2004_ex02.pdf [03/04/2007].

Biermann, F., and S. Bauer (Eds). 2005: *A World Environment Organization: Solution or Threat for Effective International Environmental Governance?* Aldershot, UK: Ashgate.

Bigg, T. 2003. The World Summit on Sustainable Development: Was it worthwhile? International Institute for Environment and Development (IIED). http://www.poptel.org.uk/iied//docs/wssd/wssdreview.pdf [03/04/2007].

Chambers, W. B. 2001. International trade law and the Kyoto Protocol: potential incompatibilities. In *Inter-linkages: the Kyoto Protocol and the International Trade and Investment Regimes*, edited by W. B. Chambers, 87–118. Tokyo, Japan: United Nations University Press.

Chasek, P. 2000. Introduction: the global environment at the dawn of a new millennium. In *The Global Environment in the Twenty-First Century: Prospects for International Cooperation*, edited by P. Chasek, 1–11. New York: The United Nations University Press.

Conca, K. 1995. Greening the United Nations: environmental organizations and the UN system. *Third World Quarterly*, 16 (3): 441–457.

D'Anieri, P. 1995. International Organizations, Environmental Cooperation, and Regime Theory. In *International Organizations and Environmental Policy*, edited by R. V. Bartlett, P. A. Kurian, and M. Malik, 153–169. Westport: Greenwood Press.

DeSombre, E. R. 2006. *Global Environmental Institutions*. London,: Routledge.

Dingwerth, K. 2005. The Democratic Legitimacy of Public-Private Rule Making: What Can We Learn from the World Commission on Dams? *Global Governance*, 11 (1): 65–83.

Downie, D. L. 1995. UNEP and the Montréal Protocol. In *International Organizations and Environmental Policy*, edited by R. V. Bartlett, P. A. Kurian, and M. Malik, 171–185. Westport: Greenwood Press.

Downie, D. L., and M. A. Levy. 2000. The UN Environment Programme at a Turning Point: Options for change. In *The Global Environment in the Twenty-First Century: Prospects for International Cooperation*, edited by P. Chasek, 355–377. New York: The United Nations University Press.

Eckersley, R. 2004. The Big Chill: The WTO and Multilateral Environmental Agreements. *Global Environmental Politics*, 4 (2): 24–40.

Eglin, R. 1998. Trade and environment. In *The Uruguay Round and beyond: Essays in Honour of Arthur Dunkel*, edited by J. Bhagwati and M. Hirsch, 251–263. Berlin, Germany: Springer.

Imber, M. F. 1996. The Environment and the United Nations. In *The Environment and International Relations*, edited by J. Vogler and M. F. Imber, 138–151. London: Routledge.

Ivanova, M. 2005. Can the Anchor Hold? Rethinking the United Nations Environment Programme for the 21st Century. *Yale F&ES Publication Series*, Report No. 7. http://environment.yale.edu/documents/downloads/o-u/report_7_unep_evaluation.pdf [03/04/2007].

Karns, M. P., and K. A. Mingst. 2004. *International Organizations: The Politics and Processes of Global Governance*. Boulder and London: Lynne Rienner.

Keohane, R. O. 1989. *International Institutions and State Power: Essays in International Relations Theory*. Boulder: Westview Press.

Pauwelyn, J. 2003. *Conflict of Norms in Public International Law: How WTO Law Relates to Other Rules of International Law.* Cambridge, UK: Cambridge University Press.

Sampson, G. P. 2002. *The World Trade Organization and Global Environmental Governance.* Tokyo, Japan: United Nations University Press.

Sampson, G. P. 2005. *The WTO and Sustainable Development.* Tokyo, Japan: United Nations University Press.

Santarius, T., H. Dalkmann, M. Steigenberger and K. Vogelpohl. 2004. *Balancing Trade and Environment: an Ecological Reform of the WTO as a Challenge in Sustainable Global Governance*, Wuppertal Paper No. 133e, Wuppertal, Germany: Wuppertal Institute for Climate, Environment and Energy. http://www.wupperinst.org/globalisierung/pdf_global/balancing_trade.pdf [03/04/2007].

Stilwell, M., and E. Tuerk. 1999. *Trade Measures and Multilateral Agreements: Resolving Uncertainty and Removing the WTO Chill Factor.* WWF International Discussion Paper, November 1999.

Streck, C. 2002. Global Public Policy Networks as Coalitions for Change. In *Global Environmental Governance. Options & Opportunities*, edited by D. C. Esty and M. H. Ivanova, 121–139. New Haven, CT: Yale School of Forestry and Environmental Studies.

Thomas, U. O. 2004. UNEP 1972–1992 and the Rio Conference. *EcoLomic Policy and Law: Journal of Trade and Environment Studies*, 4 (Special Issue).

UNDP (United Nations Development Programme). 2006. *Global Partnership for Development. UNDP Annual Report 2006.* New York, NY: UNDP. http://www.undp.org/publications/annualreport2006/english-report.pdf [03/04/2007].

UNEP (United Nations Environment Programme). 2007. *UNEP 2006 Annual Report.* Nairobi, Kenya: UNEP. http://www.unep.org/pdf/annualreport/UNEP_AR_2006_English.pdf [03/04/2007].

Vogler, J. 2007. The International Politics of Sustainable Development. In *Handbook of Sustainable Development*, edited by. G. Atkinson, S. Dietz and E. Neumayer, 430–446. Cheltenham: Edward Elgar.

Von Frantzius, I. 2004. World Summit on Sustainable Development Johannesburg 2002: A Critical Analysis and Assessment of the Outcomes. *Environmental Politics*, 13 (2): 467–473.

Wagner, L. M. 2005. A commission will lead them? The UN commission on sustainable development and UNCED follow-up. In *Global Challenges: Furthering the Multilateral Process for Sustainable Development*, edited by A. C. Kallhauge, G. Sjöstedt and E. Corell, 103–122. Sheffield: Greenleaf.

Werksman, J. 2001. Greenhouse-gas emissions trading and the WTO. In *Interlinkages: the Kyoto Protocol and the International Trade and Investment Regimes*, edited by W. B. Chambers, 153–190. Tokyo, Japan: United Nations University Press.

WRI (World Resources Institute) 2003. *World Resources 2002–2004 – Decisions for the Earth: Balance, Voice, and Power.* Report produced in collaboration with UNDP, UNEP, and the World Bank. Washington, DC: WRI.

Young, Z. 2002. *A New Green Order? The World Bank and the Politics of the Global Environment Facility.* London: Pluto Press.

Zelli, F. 2006. The World Trade Organization: Free Trade and its Environmental Impacts. In *Handbook of Globalization and the Environment*, edited by K. V. Thai, D. Rahm and J. D. Coggburn, 177–216. London, UK: Taylor & Francis.

Notes

1. Both authors contributed equally to this essay.

2. Africa (16 seats), Asia (13), Eastern Europe (six), Latin America and Caribbean (10), Western Europe and others (13)

3. These figures are based on a personal communication from UNEP and refer to its own statistics from 31 December 2005. The annual budget includes all sources of funding, demonstrating the importance of earmarked funds if compared with the size of the Environment Fund.

4. http://www.worldbank.org [03/04/2007]. This figure ranks even higher when accounting for environmental implications of other World Bank projects: while, in 2000, the Bank had officially spent US $1,830m. on projects under the label of 'environmental and natural resource management', the World Resources Institute assumes an overall portfolio of $5,000m. in environmental projects for the same year (WRI 2003: 152).

5. http://www.gefweb.org [03/04/2007].

6. Specialized agencies are autonomous organizations working with the UN through ECOSOC (http://www.un.org/aboutun/ [03/03/2007]).

7. These decisions were partly based on 'general exceptions' which two WTO agreements grant for measures protecting human, animal or plant life or conserving natural resources (Article XX GATT and Article XIV GATS [General Agreement on Trade in Services]).

8. The authors are grateful to the World Resources Institute for permission to reprint this table. It first appeared (in a longer version) as 'Table 7.1: Selected Intergovernmental Organizations that Influence Environmental Governance' in *World Resources 2002–2004 – Decisions for the Earth: Balance, Voice, and Power* (WRI 2003: 142–43). The section on the WMO has been added by the authors.

Environmental Movements[1]

BRIAN DOHERTY

Environmentalism has a good claim to be the most significant social movement of the modern era. Environmentalist ideas that were once seen as marginal or extreme are now mainstream in national and international politics; green parties have become established in many parts of the world, particularly in Western Europe where they have played a role in national governments in several countries; environmental groups have increased their support and resources in the industrialized countries to a position where governments cannot afford to ignore them; and in the global South tens of thousands of environmental groups have emerged in recent decades, often linked through transnational networks, and sometimes able to exert pressure on national governments and international bodies.

This growth in influence has been rapid, concentrated mostly in the last four decades in the North and more recent still in the South. It suggests a picture of a global movement on an ever-upward trajectory of power, but while environmental movements and groups have undoubtedly grown in strength, it would be misleading to project this with any confidence into the future. The growth of environmentalism has been accompanied by significant diversification in strategies and ideology and the differences between different types of environmental groups remain significant, as do the cross-national varieties of environmentalism (Guha and Martínez-Alier 1997), undermining the idea of a unified global environmental movement. Furthermore as environmental ideas are taken up by states and business, those who see environmental movements as too institutionalized to be able to respond to new ideas (in the USA, Brulle and McCarthy 2005; Bosso 2005) or unable to show how they will be able to transform society. (in Europe, Blühdorn 2007) have questioned its purpose.

This essay will provide an overview of the evidence and debates about environmental movements. Beginning with an examination of the historical background, it shows how environmental groups are difficult to define as a single movement. Two forms of institutionalization are examined: the transformation of green parties and the growth of the large-scale environmental movement organizations such as Greenpeace. Although some groups have institutionalized, new grassroots groups have also emerged in both North and South. This diversity means that it is no longer easy to define environmental groups as necessarily non-state actors. In many countries groups can be categorized by their position as either involved in neo-liberal forms of governance, linked to national states and international institutions, or as part of outsider networks committed to campaigning for environmental justice.

HISTORICAL BACKGROUND

The first wave of groups campaigning on environmental issues emerged in Western Europe and North America in the late 19th century. Most of them were concerned with the conservation of wildlife, landscape and biodiversity and they often framed their action as a defence of national heritage, evoking patriotic sentiments. Most were not radical, and their support was usually from the affluent and the élites. Some of the largest contemporary conservation organizations, such as the Royal Society for the Protection of Birds and the National Trust in the United Kingdom, and the Sierra Club in the USA, have their roots in this period. In the USA the main division was between those such as Gifford Pinchot who saw conservation as a means to a more rational management of natural resources such as forestry, and John Muir, the founder of the Sierra Club, who argued for the preservation of areas of wilderness unspoilt by human civilization.

These groups were, of course, not the only ones engaged in environmental conflicts. As industrial production grew, resource extraction produced many conflicts between expanding capital and labour, and also between colonizer and colonized. Although groups that opposed them did not define themselves as environmentalists, pollution from mining, and deforestation had major environmental impacts on colonized societies (Martínez Alier 2002). The expansion of European settlement and colonization provoked some from the colonial countries to a new critique of the effects of colonialism on the natural environment, one that drew in part on indigenous knowledge (Grove 1995). For the most part, however, the colonial territories were regarded as sources of natural and human resources for the metropolitan powers and not until late in the 20th century were they also recognized as central to global environmental solutions.

The second wave of environmentalism emerged as a new sense of global politics was unfolding in the 1960s. The US defeat in Viet Nam and new guerrilla movements in Latin America and Africa seemed briefly to open the possibility of new kinds of socialism, even within a geopolitics that was dominated by the Cold War. In the West, the New Left and student movements were challenging the post-war consensus that all the major ideological questions had been settled so that politics was only about rational management of economic growth. The new environmental groups that developed in the late 1960s and 1970s in Northern countries were influenced by this context. Whereas the conservation groups had mainly relied on élite support and insider lobbying, new groups, including Greenpeace and Friends of the Earth (FoE), used public protest, were openly critical of governments, and were part of a new green movement committed to the need to create an alternative society (Doherty 2002).

New evidence about the global pressures on limited resources was also at the centre of the new environmentalism. The UN Conference on the Human Environment in Stockholm (1972) and the Club of Rome's *Limits to Growth* report (1973) reinforced the new global agenda for environmentalism. This was often split between the positive potential for a new kind of politics, based on participatory and local democracy in an egalitarian and redistributive society, and the apocalyptic dangers of collapse due to exhaustion of resources. The politics of population growth often served as a dividing point between the emancipatory and the authoritarian tendencies in the new environmentalism. For environmental authoritarians, population growth in the South threatened to exhaust finite resources, led to the 'lifeboat ethic', and potentially to the need to restrict freedoms in order to prevent a collapse of society. For some, democracy was too much of a risk, since people were unlikely to vote for the restrictions on consumption seen as necessary to avert crisis (Ophuls 1977). In contrast, for other environmentalists, it was less population growth that was the problem than inequality (Kemp and Wall 1990). High rates of population growth had been a feature of capitalist development in the wealthiest countries so it was misleading to see this as a 'Southern' problem. For emancipatory environmentalists the main issue was the high rate of consumption in the North—with 20% of the world's population controlling 80% of its resources. For this group ecology and the limits to growth provided arguments for a global redistribution of wealth and a reduction in consumption by the rich in order to provide conditions for meaningful positive freedom for the majority of the world's population, which required a deeper democracy and a new global politics.

Divisions over questions of social justice and democracy continue to be evident in environmental movements, but they have rarely been visible publicly in the main environmental groups.[2] Since the early 1970s many conservation groups have moved to a more international focus; the Worldwide Fund for Nature for instance moved from a narrow stress on conservation to an emphasis on sustainable development as the most effective means to protect wildlife (Rootes 2006). Involving human communities in conservation strategies is the predominant strategy for conservation organizations in the 21st century, even if its implementation is fraught with controversy because of potential conflicts between the interests of human communities and 'other nature' (Agrawal and Gibson 2001).

TYPES OF ENVIRONMENTAL GROUPS

Environmental groups can be categorized in a number of ways. Divisions over issue focus, and ideology between conservation groups and political ecology groups are still evident, but in many cases are less sharp than they were in the 1970s. As some of the larger conservation groups have broadened their remit to include sustainable development, some of the political ecology groups founded in the 1970s have become more institutionalized, developing larger bureaucracies with specialist staff and engaging in insider lobbying and education as well as outsider pressure on government. Another way of classifying environmental groups is to look at their primary emphasis. Doherty and Doyle (2006) set out a threefold division between post-material, post-industrial and post-colonial environmental movements. Post-material movements concentrate on the rights of 'other nature', conservation of species and preservation of wilderness. Post-industrial movements challenge the excesses of the industrialist project; the rights of corporations to pollute and degrade and the dwindling of the earth's resources as they are fed into advanced industrial machines. Post-colonial movements cast green concerns in the narrative of the colonizer versus the colonized, the dichotomous world of affluence and poverty. Examples of movements using these three frames can be found in all parts of the world, but post-material movements have been dominant in the USA and Australia and post-colonial movements in the South. Post-industrialism has been strongest in Europe, but there both post-material and post-colonial narratives have also been part of environmentalist discourse and from an early stage, as for instance in René Dumont's 1974 campaign for the French presidency.

Organizationally and strategically, environmental groups can also be divided into different types. Diani and Donati's typology (1999: 16) is useful in locating groups according to how they respond to two requirements, resource mobilization and political efficacy. The main resources available to groups are either professional expertise, for which money is needed, or the time of activists. Groups that concentrate on professional resources have to raise finance either through mass membership or other sources such as grants from government or foundations or corporate donations. This kind of approach entails a particular kind of bureaucracy able to generate income and with specialist sub-divisions. Groups that concentrate on using the time of activists need to develop a structure that enables activists to feel that their action is effective and influences what the organization does. In terms of the most effective political means of action, groups can be divided between those that give priority to challenging and disruptive action and those that mostly use conventional pressure.

	Conventional Pressure	Disruption
Professional resources	*Public interest lobby*	*Professional protest organisation*
Participatory resources	*Participatory pressure group*	*Participatory protest organisation*

Table 1 Four Ideal-Types of Environmental Organisations (from Diani and Donati 1999: 16).

As Table 1 shows, this produces four ideal-types of environmental group. The public interest lobby is managed by a professional staff, has low participation and uses conventional pressure tactics. Many of the larger US environmental organizations such as the Nature Conservancy are of this kind. The professional protest organization combines the mobilization of finance from supporters and donors to support a professional staff with the use of direct action and disruption, though sometimes alongside more conventional tactics. It has minimal opportunities for participation by supporters in determining the organization's strategy. Greenpeace is the best example of this kind of group. The participatory pressure group uses conventional pressure group lobbying and non-confrontational tactics and has a structure that allows members to influence policy. The Sierra Club in the USA, and Bund für Umwelt und Naturschutz Deutschland (BUND/FoE) in Germany are groups that fit this type. The participatory protest organization uses direct action and usually has little or no central organization. Radical networks such as Earth First! (Wall 1999), parts of the anti-nuclear movement (Rüdig 1990; Epstein 1991) and some environmental justice groups in the USA fit this type (Schlosberg 1999).

INSTITUTIONALIZATION

Diani and Donati's typology was developed with Western Europe in mind where there has been a widespread trend towards institutionalization in environmental movement organizations. In the USA environmental organizations were mostly established as initiatives of foundations or at the behest of state agencies rather than grassroots organizations and some only became membership organizations when federal funds diminished during the Reagan Administration (Bosso 2005). They did not, however, change their strategy in response. The disconnection between the Washington-based and lobby-focused staff of the major environmental movement organizations and their rank and file supporters has been the subject of critical commentary from those who see their unwillingness to take political action that might reduce their access to decision-makers as a failure to challenge the main-stream (Dowie 1995; Dryzek et al. 2003; Brulle and McCarthy 2005). Thus a vital question in assessing Northern environmental movements is whether institutio-nalization, whether longstanding or recent, entails deradicalization.

When we speak of the institutionalization of the environmental movement we combine a number of phenomena that are analytically distinct. Van der Heijden (1997) distinguishes between organizational growth, the increase in membership and income of major groups; internal institutionalization, the professionalization and centralization of the organization; and external institutionalization, which includes a shift from disruptive forms of protest to lobbying and more conven-tional protest action. This can also involve greater ties with government, inter-national institutions and business.

The growth in membership of environmental groups has been fairly consistent in Europe, Australia and the USA since the early 1970s, but there have been several periods of remarkable growth, often coinciding with periods when environmental issues were politically prominent, such as the late 1980s and the early 2000s. It is often the older conservation organizations that have grown most steadily. Greenpeace and some national Friends of the Earth groups saw sig-nificant declines in the mid-1990s, and for them this is often more serious than for most of the conservation groups because they avoid corporate donors and are therefore reliant on members and supporters for their finances. Greenpeace had to cut its staff and seek extra money from a minority of its supporters in order to offset the loss of resources from falling numbers of supporters in the USA. Greenpeace world-wide has, however, been able to stabilize and sustain its income despite this instability without having to seek corporate or government funds. Organizational growth provided these groups with the option to invest in specialist expertise, expand their research and so to play a more effective role in policy debates

The emergence and consolidation of green parties in Western Europe in the 1990s and in Australasia could also be seen as an example of the institutionalization of environmental movements. Green parties began to gain seats in national parliaments in the mid-1980s and became the first new party family to emerge in Western Europe in the post-war period. Green success was very much dependent upon the existence of some degree of proportional representation in national electoral systems, hence it was in Austria, Belgium, Finland, Germany, Ireland, Italy and Switzerland that greens first gained elected representatives. In countries such as the United Kingdom and France greens were restricted to European or regional parliaments, unless they were able to form an electoral alliance, as the French Greens did with the Socialists in the 1997 and subsequent elections. Greens have been part of governing coalitions in Belgium, Finland, France, Germany, Ireland and Italy but always as a minority partner, and often only able to exert weak influence on policy (Rihoux and Rüdig 2006). The greens have been useful allies on occasions for environmental groups, but the relationship between green parties and environmental groups has not always been as close as might be expected. Most environmental groups adopt a non-partisan political position, which means that few endorse the green party specifically, because they regard it as important to retain access to other parties, particularly the larger governing parties, and also because many groups have supporters who are voters for other parties. While some green party representatives have experience in environmental organizations they also drew on support from a wider network of social movements, including peace movements, women's movements, and minority and migrant workers' groups. The campaigns against nuclear energy in the 1970s and 1980s were formative for many Western European green parties. These campaigns also had strong support from parts of the New Left, and this broader social movement experience largely explains the combination of left-wing and environmentalist themes in green party programmes.

Green parties themselves have undergone processes of institutionalization, becoming reconciled to the limits of their own power as parties with a support base of usually 5%–10% of the voters, and the need to maximize their chances of being in government. Internally, both green parties and larger environmental organizations have been accused of moving away from a model of grassroots politics, in which the members had power over leaders, to a more passive form of participation, often referred to as 'chequebook activism' in which most members are happy to provide financial support but less willing to give up their time for more demanding forms of activism.

For environmental radicals all three kinds of institutionalization—organizational growth, internal and external—lead to deradicalization. Green parties set out to be new kinds of parties, with power concentrated at the grassroots and strong constraints on their leaders, including restrictions on periods in office, the rotation of offices, a weak party bureaucracy and a decentralized decision-making structure. Many of these measures were abandoned, however, because they conflicted with the need to compete effectively in elections and to act professionally in national parliaments. In that sense, green parties are less radical than they mostly set out to be. However, it is not always clear that this is what has happened in environmental movements. While there has been a trend towards internal institutionalization among many of the newer environmental organizations formed in the 1970s, such as the European branches of FoE and Greenpeace, this has not necessarily resulted in an overall deradicalization of the movement. First, because some of the larger conservation groups were never radical in their strategies and, as noted, some have broadened their issue focus to include questions of development, even if they remain relatively moderate politically. Second, if institutionalization was deradicalizing the green movement, we would expect to see a decline in protest, but it is not clear that environmental protest has declined, even if institutionalization has increased. Patterns vary considerably between countries, and the data is far from comprehensive, but in the most systematic analysis of environmental protest in Europe (Rootes 2003), some counties such as Germany and the United Kingdom maintained high levels of disruptive protest action in the 10 years from 1988–97, a period when instutionalization had taken hold in most large environmental movement organizations. In Germany and the United Kingdom parts of the movement had remained uninstitutionalized, notably the anti-nuclear energy protesters in Germany, and in the United Kingdom new protest coalitions against road building combining sub-cultural radicals committed to direct action and more conventional local environmental campaigns had developed independently of the institutionalized environmental organizations. In France too, there has been a sudden and unexpected emergence of new direct action coalitions on issues such as genetically modified crops, even as environmental groups are growing in membership (Hayes 2007). Thus, it is important to remember that the formally organized and institutionalized groups are not equivalent to the whole movement; informal and uninstitutionalized groups can continue to exist alongside the large environmental organizations.

Institutionalization can sometimes constrain the ability of the larger groups to take action, since they may be vulnerable to prosecutions leading to fines that threaten their financial viability and therefore need to give priority to protecting the long-term interests of the organization and staff. But other parts of the environmental movement can often fill the gap left by more cautious institutionalized groups, while the latter are able to provide them with research or to use their access to élites to exert pressure that is not visible to the public. So, even when organizational growth creates constraints, it is not clear that the end result is necessarily a deradicalization of the movement as a whole.

ENVIRONMENTAL RADICALS: DIRECT ACTION AND ENVIRONMENTAL JUSTICE NETWORKS

Not all environmentalists are moderate in the Northern countries. In recent decades the resurgence of direct action protests in several countries and the development of environmental justice networks in the USA has qualified the view that all environmental groups are now mainstream and moderate.

Environmental direct action (EDA) groups are difficult to classify because they often lack any formal organization or professed policy beyond some very general commitments to take responsibility for defending the environment using (usually) non-violent forms of direct action. Direct action refers to action that is potentially illegal, and designed to obstruct or disrupt the actions of governments or other opponents who are deemed to be acting unjustly. This can include forms of civil disobedience in which symbolic illegal acts are taken as an appeal to the wider public or to the government on a particular issue. This is the strategy used by Greenpeace and is very dependent on successful media coverage. EDA groups usually go beyond civil disobedience, however. Civil disobedience entails a willingness to accept the consequences of arrest and seeking to distinguish the particular illegal action from a wider challenge to the principles of the law, but direct action groups are usually influenced by an anarchistic sub-culture which is critical of the limits of parliamentary, as opposed to direct, democracy, seeks to challenge hierarchies and gives priority to individual responsibility and con-science over political obligation. This means that most such activists view the question of where and when to take illegal action as a pragmatic one, governed by what is most likely to be effective and feasible, rather than a moral one. There is, however, widespread consensus on commitments to non-violence in these net-works if non-violence is defined as avoiding harm to human and non-human animals. It follows that damage to property is not seen as violent. Even if that view of violence is arguable, it is clear that there is a significant moral difference between violence to people and sentient animals and damage to inanimate objects. Analysis of UK EDA groups shows that there was minimal violence in their protests (Rootes 2003; Plows et al. 2003; Doherty 2007). The 'violence' of EDA has nevertheless been controversial on occasions. The FBI has pursued some US EDA activists as 'terrorists' and some have had draconian sentences passed for acts of arson against SUVs and buildings that were deemed 'terrorist' acts by the courts.

EDA groups have emerged to varying degrees in most Northern countries. They have perhaps been strongest in the USA and Australia where their main focus was on defending wilderness, and the United Kingdom, Germany and France, where groups often linked their campaigns against nuclear energy, new roads, airports or GM foods more directly to questions of social justice. While there were few direct examples of successful outcomes of direct action protests, there was considerable evidence of public support, and direct action pushed issues that were marginal to the forefront of public debate (Dunleavy et al. 2005). Nevertheless direct activists were always a small part of the movement overall and because of their radicalism they often had difficult relationships with the mainstream environmental groups, even if there was in practice sometimes a division of labour between them. Also, EDA is difficult to sustain over long periods. It relies on intense commitment and usually the adoption of a whole alternative way of life. This helps to explain why it is not a mass movement, but also why relatively small numbers of activists linked by strong bonds were able to sustain high intensity action for years at a time.

Like the EDA movements, the US environmental justice (EJ) movements also emerged in part in reaction against the limitations of the established environmental organizations, which were seen as too oriented to the protection of the natural environment and blind to environmental impacts on people. In the late 1970s evidence emerged that toxic waste dumps were damaging the health of many in disadvantaged communities. The inspiration provided by the high-profile campaign by women residents of Love Canal, New York, led by Lois Gibbs, in protest against the failure of authorities to accept responsibility for building housing and a school over a toxic waste dump, and the residents of Warren County, North Carolina, against toxic dumping led existing and other, new community and civil rights groups to take up the issue and to the creation in 1980 of a Federal 'superfund' to clean up toxic waste sites. The issue of racial injustice became central during the 1980s as evidence accumulated that communities of color were much more likely to be exposed to environmental hazards than white communities. Many environmental justice groups grew out of civil rights networks in local communities and were markedly different in ideology, political style, language and organization from the institutionalized environmental movement (Pellow and Brulle 2005). Links between environmental justice groups grew like rhizomes, from below without any strong central organization (Schlosberg 1999). The major success of the environmental justice movement has been to connect issues of environment with issues of inequality in the USA. There has been some institutionalization of the environmental justice networks at the national level in recent years, but the frame of environmental justice has been adopted by many groups fighting environmentally damaging and socially unjust development projects in the South, where the same issue of disproportionate exposure of the poor to hazards applies.

ENVIRONMENTALISM IN THE MAJORITY WORLD

Environmental movements have often been perceived as a luxury only available to the wealthiest countries. The growth of environmentalism in Australasia, Western Europe and North America after the 1960s was explained by Ronald Inglehart (1977, 1990) as part of a silent revolution in which post-material values emerged as an alternative to the previously unchallenged material values of economic growth and physical security. Inglehart argued that the post-war generations could afford to consider quality-of-life issues such as the environment because they had been socialized in conditions of affluence and security. This also seemed to be borne out by the evidence about the activists in the environmental movement, who came mainly from new middle-class professions and had above-average education. The post-material value hypothesis applies well to Northern green activists, not least because in surveys post-materialism is strongly linked with higher education, itself a significant predictor of political activism. It is less useful, though, for explaining the wider support for green issues in the North, predicting activism on questions of environmental justice, and involvement in environmental struggles in the South, all of which involve other social groups than the higher-educated post-materialists.

Rootes (2005) argues that concern with global green issues and post-material concern with the rights of 'other nature' tend to be concentrated among those with higher education, while concern with environmental risks and hazards is over-represented among those with less formal education, because those with more education have more confidence and analytical skills to assess risks. Conversely, those with less education are more likely to be in poorer communities with higher levels of exposure to pollution. This is particularly the case in Southern countries where many environmental battles are intense local struggles to protect the means of livelihood and life chances of local communities against the threat of their destruction as a result of development projects or the extraction of natural resources for export. In the South the ability to mobilize on environmental issues is dependent upon the strength of democracy, the protection for civil liberties and the possibility of using the courts to act against opponents. Despite the wave of democratization since the 1980s, these rights remain fragile and vulnerable in most countries of the South and non-existent in others, so that struggles on environmental issues are also struggles for democracy and against the unequal distribution of power.

Nevertheless, this broad generalization hides huge disparities between countries with long-established democracies such as India, transitional and relatively affluent societies in East Asia and parts of Latin America, where democracy has only recently been either established, or re-established, and others, such as most of sub-Saharan Africa and parts of South East Asia, Central Asia and the Middle East, where democracy does not exist in any meaningful sense. In states with strong industrial bases, such as South Korea, Taiwan, and in Eastern Europe before 1989, authoritarian rulers often tolerated activism on specific environmental issues in the pre-democratic period because this did not offer a direct challenge to the legitimacy of the political system. In these countries environmentalism helped in the democratization process, but once the transition was achieved, environmental movements lost much of their support as the new political élites focused on economic development and catching up with the wealthiest countries. It is possible that a similar trajectory will be followed in China, where the environmental costs of rapid economic growth have begun to provoke local protest, but as in other non-democratic states, the possibilities for building environmental movement organizations remain heavily constrained and dependent on the support of leading figures in the political élites.

In the South then much environmentalism is a struggle to resist hazards and loss of livelihood. Dwivedi (2001) cautions us, however, not to assume that the local politics of the defence of livelihoods necessarily sums up all Southern environmentalism. With reference to Indian environmental conflicts, he challenges the idea that there is always a dichotomy between localized environmentalism based on local cultural values and other forms of action and knowledge. Rather, many environmental conflicts are embedded in a more complex local-global nexus, involving network ties between local communities, travelling radical activists, national environmental organizations and international supporters. Local communities are able to fashion contacts with others to their own ends and able to link local cultural values and dominant forms of scientific and economic knowledge. Similar ties connect local, national and global levels in other prominent cases, such as the conflicts over the degradation of the Niger delta by the oil industry. Most conflicts over environmental issues in the South are local, but those that achieve prominence often move to an international level, with environmental groups seeking to exploit what has been termed 'the boomerang effect' (Sikkink 2005), in which environmental groups in Southern countries work with allies in Northern countries, who are able to use their access to transnational financial institutions or to their own governments to put pressure on the authorities or companies in the Southern country. This kind of transnational advocacy network has grown as the costs of international communication and travel have reduced. However, the danger is that by moving to the transnational level, local agendas get displaced by the need to work within the criteria of international institutions, or to the priorities of transnational networks (Widener 2007). Since transnational environmental action of this kind has increased, it is worth comparing two of the most important activist-oriented transnational environmental groups.

TRANSNATIONAL ENVIRONMENTALISM

Greenpeace and FoE are the two most significant of the newer environmental organizations with an international focus but they differ significantly in their resources, organization and strategy. In 2004 Greenpeace had branches in 38 countries, while FoE had affiliated groups in 71 countries. Greenpeace had 2.7m. identified supporters, while FoE had around 1.5m. members and supporters with around 5,000 groups world-wide. Both are transnational environmental organizations, but while FoE is decentralized, Greenpeace is highly centralized, with an international secretariat in Amsterdam, which controls a significant proportion of Greenpeace's total income (€39m. out of a total for Greenpeace world-wide of €159m. in 2004). Greenpeace International establishes branches in countries rather than existing groups applying to join Greenpeace. Friends of the Earth International (FoEI) is, in contrast, organized federally, with each country having an equal vote and new applicants having to prove their record of campaigning to the whole membership in an extended process before they can join. FoEI has a detailed set of democratic processes which it uses to develop common strategies and ideas among its diverse membership, but this emphasis on participation and dialogue can produce tensions when different perspectives clash, and so much effort is devoted at the international level to building trust and solidarity. FoEI has as a result been more open to influence from Southern groups which have helped to position it as a radical actor at the international level, giving priority to opposition to neo-liberalism, support for grassroots mobilization and resistance, and a faith in radical green alternatives. Greenpeace, in contrast, concentrates on relatively few and specific campaigns at the international level, and avoids slogans or issue frames that sound ideological.

While FoEI is better able to claim to have integrated diverse forms of radical environmentalism from North and South, its international arm lacks the resources of Greenpeace, or a presence in some important states. Because it refuses to create franchises, unlike Greenpeace, FoEI has no groups in Russia, India or China, and the income of its international organization was only €1.5m. in 2004. This compared to the £5m. income of FoE England Wales and Northern Ireland. Most national FoE groups fail to give the 1% of their income that they are supposed to provide to the international organization and many FoEI member groups hold on to their original names and do not identify themselves as FoE groups in their own countries. Thus while some FoE groups in the North and South are very committed to joint international campaigns, others are more reluctant to commit themselves, either because it is too radical for some in the North, or too conservative for some in the South. FoEI is a complex network: it is one of the few spaces where multiple forms of environmentalism interact, but it also shows the difficulties facing the development of global environmental organizations (Doherty 2006).

Greenpeace and FoEI are outsiders in international negotiations, rarely present in the invited spaces provided by intergovernmental institutions or international financial institutions. They have nevertheless been able to influence outcomes through outside pressure on occasion, as with the extractive industries review by the World Bank in 2001, in which FoEI, along with other organizations, helped to move the Bank away from support for mining, gas and oil projects on social justice and environmental grounds. Other groups, such as WWF, the Wildlife Conservation Society and Conservation International in the USA, are more often part of governance networks, linking NGOs, international institutions and national governments. These networks raise major issues of accountability and representation in the South, since they are often acting to influence access to aid or development budgets without any clear claim to represent those in the South most affected by their decisions (Duffy 2006). This split between governance environmental groups and outsider groups, often with a stronger commitment to an emancipatory critique of existing power relations, is important because it shows that the environmental movement remains differentiated by type and location and undermines easy generalizations about the environmental movement as either a constituent of a new global civil society, or acting only on behalf of Northern post-material agendas. Parts of the environmental movement can be shown to be acting in these ways, but not the movement as a whole.

POLITICAL CONTEXTS—OPPORTUNITIES AND OUTCOMES

Analysis of the outcomes of social movements is bedevilled by a number of obstacles. First, movements rarely see their demands translated directly into public policy. And when they do, the number of intervening variables is so substantial that it is very difficult to establish a causal relationship between movement action and policy outcome (Amenta and Caren 2004; Giugni 2004). For instance, Britain has moved further and faster than some countries with weak environmental movements to develop policy on climate change, but while there may be a correlation between strength of environmental movement and willingness to act, there are other possible causal factors, such as the response of other interest groups, public opinion, competition between the political parties and international pressures. It would be unconvincing to say that the environmental movement had played no role in shaping policy on climate change, but hard to show that it could claim sole credit.

Most analysts argue that policy impact is facilitated best by splits in élites, because this can provide allies for the movement within the polity. Tarrow (1998) argues that it is when the costs of action are lowered because repression reduces or new powerful allies become available that social movements are most likely to take action. These political opportunities help to explain why movements emerge when they do. This kind of model works best for explaining sudden large-scale cycles of protest in countries or areas with a tradition of strong repression of public protest, such as the Philippines in 1986 or Eastern Europe in 1989, but many environmental groups emerge outside major cycles of protest, either in reaction against the limits of existing organizations, as with the direct action groups, or because of new evidence about environmental hazards, in the case of environmental justice groups in the USA and the South. With the expansion of the environmental movement transnationally, new groups also emerge in response to the incentives provided for NGOs in new networks of governance, nationally and internationally. This is particularly relevant for new NGOs from the South. Thus the causes of the emergence of environmental movements vary considerably.

A further factor mediating the impact of environmental movements is how far their goals touch on the core interests of the state. It can be argued that environmentalism has always lost out in confronting the core imperative of economic growth because this is so central to the purpose of the state in capitalist societies. But it is possible that a new state imperative of environmental conservation is emerging around both the politics of climate change and of manufactured risks such as GM foods, which might establish movements' goals as a new state imperative alongside and in tension with economic growth. In an analysis of the relationship between environmental movements and the state in the North, Dryzek et al. (2003) argue that relatively open political systems that responded early to environmentalism are not necessarily conducive to the development of the strong environmental movements required to achieve this end. Movements that lack a radical protest constituency lack the ability to mobilize pressure and also lack the internal movement debates necessary to challenge the orthodoxies of policy, including the unquestioned commitment to economic growth.

CONCLUSION

While the central political status of environmental problems in the early years of the 21st century seems to indicate that environmentalism has come a long way, it has also been argued that environmental movements have essentially run out of steam. Writing about the politics of 'the advanced consumer democracies', Blühdorn (2007) argues that the dominant *Zeitgeist* is based on a societal self-deception in which it is acknowledged that radical changes are needed in order to overcome the problems of unsustainability, but there is no confidence, even among environmentalists, that there is any alternative to the current system. Blühdorn calls this the post-ecologist condition and described in this way it is a plausible description of some forms of environmentalism, particularly those that have become enmeshed in practices of governance. Environmental movements are too diverse for this to apply to all of them, however, and this diagnosis underplays the consequences of their differentiation over time and the influence of environmentalisms in the global South. In the 1970s very few of the hard questions being asked by environmentalists about how we should live had an audience other than fellow environmentalists. Now, it is not that environmentalists no longer ask those questions, but they ask so many of them and in so many different voices that it is impossible to define a single environmental agenda.

It is also wrong to see it as the responsibility of an as yet non-existent global environmental movement to develop a single new model of society. One of the defining features of contemporary social movement politics is a greater awareness of the value of diversity. The global justice movement in social forums, in networks such as People's Global Action, and parts of the environmental movement, including FoE, has worked to create spaces in which debate and dialogue can develop between actors with different perspectives on how to achieve a socially just sustainable society. Doug Torgerson (2006) has called this a green public sphere, emphasizing recognition of the need for greens to agree to disagree because the profound differences that structure global politics mean that it is impossible to conceive of a single global form of environmentalism. This is a more accurate assessment of the current situation of environmentalism and one that reflects four decades of rapid and uneven growth in very different national contexts.

Bibliography

Agrawal, A. and C. C. Gibson (Eds). 2001. *Communities and the Environment.* New Brunswick: Rutgers University Press.

Amenta, E., and N. Caren. 2004. The Legislative, Organizational, and Beneficiary Consequences of State-Oriented Challengers. In *The Blackwell Companion to Social Movements*, Oxford: Blackwell, pp. 461–488, edited by D. A. Snow, S. A. Soule and H. Kriesi.

Blühdorn, I. 2007. Sustaining the Unsustainable: Symbolic Politics and the Politics of Simulation. *Environmental Politics,* 16 (2): 251–275.

Bosso, C. J. 2005. *Environment, Inc. From Grassroots to Beltway.* Lawrence, KS: University of Kansas Press.

Brulle, R. J., and J. Craig Jenkins. 2005. 'Decline or Transition: Discourse and Strategy in the U.S. Environmental Movement', presented at a special session organized by the Environment and Technology Section, American Sociological Association Conference, Philadelphia, PA, August.

Diani, M., and P. Donati. 1999. 'Organisational Change in Western European Environmental Groups: A Framework for Analysis'. *Environmental Politics*, 8, (1): 13–34.

Doherty, B. 2002. *Ideas and Actions in the Green Movement.* London: Routledge.

Doherty, B. 2006. 'Friends of the Earth International: Negotiating a North-South Identity', *Environmental Politics*, 15 (5): 860–880.

Doherty, B. 2007. Environmental Direct Action Protests in Manchester, Oxford and North Wales: A Protest Event Analysis. *Environmental Politics*, 16 (5), forthcoming.

Doherty, B., and T. Doyle. 2006. Beyond Borders: Transnational Politics, Social Movements and Modern Environmentalisms. *Environmental Politics*, 15, (5): 697–712.

Dryzek, J. S., D. Downes, C. Hunold and D. Schlosberg. 2003. *Green States and Social Movements: Environmentalism in the United States, United Kingdom, Germany and Norway.* Oxford: Oxford University Press.

Dunleavy, P., H. Margetts, T. Smith and S. Weir. 2005. *Voices of the People: Popular attitudes to democratic renewal in Britain*, London: Politico's.

Dwivedi, R. 2001. Environmental Movements in the global south: outline of a critique of the 'livelihood' approach, 227–247. In *Globalization and Social Movements*, edited by P. Hamel, H. Lustiger-Thaler, J. Pieterese and S. Roseneil. Basingstoke: Palgrave.

Epstein, B. 1991. *Political Protest and Cultural Revolution: Nonviolent Direct Action in the 1970s and 1980s.* Berkeley: University of California Press.

Giugni, M. G. 2004. *Social Protest and Policy Change: Ecology, Antinulcear and Peace Movements in Comparative Perspective.* Lanham: Rowman and Littlefield.

Grove. R. H. 1995. *Green Imperialism: Colonial Expansion, Tropical Island Edens and the Origins of Environmentalism, 1600–1860.* Cambridge: Cambridge University Press.

Guha, R., and J. Martínez-Alier. 1997. *Varieties of Environmentalism: Essays North and South.* London: Earthscan.

Hayes, G. 2007. Collective Action and Civil Disobedience: The Anti-GMO Campaign of the Faucheurs Volontaires, *French Politics,* 5 (3), forthcoming.

Kemp, P., and D. Wall. 1990. *A Green Manifesto.* Harmondsworth: Penguin.

Martínez-Alier, J. 2002. *The Environmentalism of the Poor: A Study of Ecological Conflicts and Valuation.* Cheltenham: Edward Elgar.

Ophuls, W. 1977. *Ecology and the Politics of Scarcity.* New York: Freeman.

Pellow, D. N., and R. J. Brulle (Eds). 2005. *Power, Justice and the Environment: A Critical Appraisal of the Environmental Justice Movement.* Cambridge, MA.

Plows, A., D. Wall and B. Doherty. 2004. Covert Repertoires: Ecotage in the UK. *Social Movement Studies,* 3 (2): 199–219.

Rihoux, B., and W. Rüdig. 2006. Analysing Greens in power: Setting the agenda, *European Journal of Political Research,* Vol. 45: 1–33.

Rootes, C. (Ed.). 2003. *Environmental Protest in Western Europe.* Oxford: Oxford University Press.

Rootes, C. 2004. Environmental Movements. In D. A. Snow, S. A. Soule and H. Kriesi (Eds), 608–640. *The Blackwell Companion to Social Movements,* Oxford: Blackwell.

Rootes, C. 2006. 'Facing South? British environmental movement organisations and the challenge of globalisation', *Environmental Politics* 15 (5): 768–786.

Rüdig, W. 1990. *Anti-nuclear movements: a world survey of opposition to nuclear energy.* Harlow: Longman.

Rüdig, W. 2006. Is government good for Greens? Comparing the electoral effects of government participation in Western and East-Central Europe. *European Journal of Political Research,* 45: 127–154.

Schlosberg, D. 1999. *Environmental Justice and the New Pluralism,* Oxford: Oxford University Press.

Sikkink, K. 2005. Patterns of Dynamic Multilevel Governance and the Insider-Outsider Coalition. In D. della Porta and S. Tarrow (Eds), 151–174. *Transnational Protest and Global Activism.* Oxford: Rowman and Littlefield.

Tarrow, S. 1998. *Power in Movement.* Cambridge: Cambridge University Press.

Torgerson, D. 2006. Expanding the Green Public Sphere: Post-Colonial Connections. *Environmental Politics,* 15 (5): 713–730.

Van der Heijden. H. A. 1997. Political opportunity structure and the institutionalisation of the environmental movement. *Environmental Politics* 6 (4): 25–50.

Widener, P. 2007. Benefits and Burdens of Transnational Campaigns: A Comparison of Four Oil Struggles in Ecuador. *Mobilization,* 12, 1: 21–36.

Notes

1. I am grateful to Chukwumerije Okereke and Christopher Rootes for comments on the first draft of this chapter.
2. An exception is the Sierra Club in the USA where arguments over whether the Club should support immigration controls due to pressures of overpopulation on the US environment raged for over a decade from the mid-1990s, although the supporters of immigration controls have thus far been defeated.

Environmental Politics in Multi-level Governance Systems

DAVID BENSON AND ANDREW JORDAN

INTRODUCTION

In the last 30 years, environmental policy-making has become considerably more multi-levelled, with many more actors and preferences to accommodate in the production and implementation of new policies. No term encapsulates this growing trend more clearly than that of 'multi-level governance' (or MLG) (Jordan 1999; 2002). This term builds on the observation that policy-making can no longer be neatly divided into national and international levels of governing, while at the same time drawing attention to the need to understand the deepening interactions between these levels in the process of governing society. Although now used in a number of contexts, MLG is still most strongly associated with the European Union (EU), where the merging (or, as some have argued, fusion) of the national and supranational levels, has arguably proceeded further than in any other setting (Hix 1998).

In recent years, scholars have found that the language, models and concepts associated with the development of MLG fit the development of EU environmental policy rather well. Indeed, due to its 'dispersed decision-making competences and the involvement of state and non-state actors it would seem to provide an excellent vehicle for analysing multi-level governance in general' (Fairbrass and Jordan 2004: 148). This claim directly contradicts that made by scholars such as Moravcsik (1998), who have steadfastly maintained that in the EU—as in other domains of supranational policy making—it is still important to distinguish between national and European policy-making. Advocates of an MLG approach, by contrast, argue that these differences have blurred to the point of insignificance.

In this essay, we explore the sites of political contestation in complex multi-levelled governance systems. In so doing, we seek to add flesh (or empirical validation) to the concept of MLG, which has been criticized for being more akin to a 'metaphor' than a theory (Rosamond 2000: 197). To give our analysis a sharper focus, we take the EU as our main point of departure. The first site emerges around the allocation of responsibilities between levels. MLG anticipates governance to be multi-level and multi-actor, but which level actually does what within EU environmental policy-making? As outlined below, this deceptively simple question is not easy to answer—there remain deep conflicts over how policy 'tasks' or powers are allocated and to whom (Benson and Jordan 2008a, b).

The second site concerns the interaction between levels, most notably the impact of the higher level (namely the EU) on lower levels (namely the national and sub-national levels). Here, we will draw upon evidence which is emerging in relation to the debate about the Europeanization of national systems (for example, Bache and Jordan 2006). The third site of politics emerges in relation to the choice of policy instruments. The MLG literature often assumes a steady shift has taken place from regulatory to non-regulatory forms of societal steering. But to what extent do we see this occurring in the domain of environmental policy?

In the remainder of this essay, we explore these three sites of contestation in greater detail, namely: *policy formulation* (what level does what?); *policy interaction* (the Europeanization of national policy in a system of MLG); and *policy instrumentation* (i.e. which instruments to use to attain policy goals). But first— and by way of introduction—we explain what is meant by the terms governance and MLG. Then we relate these to the ways in which the EU has evolved, and indeed what it has evolved into, focusing specifically on the development of its environmental powers.

MAPPING THE CONTOURS OF EU ENVIRONMENTAL GOVERNANCE

(Multi-level) governance: an introduction

Governance is definitely a term in good currency. But it has also been described as a 'notoriously slippery' one (Pierre and Peters 2000: 7); multiple meanings and competing definitions abound (Kohler-Koch and Rittberger 2006: 28). Hirst (2000), for example, offers five different interpretations, Rhodes (1996) six, and van Kersbergen and van Waarden (2004) no fewer than nine! Kooiman (2003: 5) understandably concludes that '[w]e are still in a period of creative disorder concerning governance'.

Generally speaking, four broad themes in the governance literature are discernible (Jordan et al. 2005: 203–204). The first is that there has been a 'putative shift from government to governance' due to a variety of exogenous and endogenous pressures, such as globalization (ibid.). On this argument, *governance* can be distinguished from *government*. Whereas the latter stresses 'hierarchical decision-making structures and the centrality of public actors ... the former denotes the participation of public and private actors, as well as non-hierarchical forms of decision-making' (Kohler-Koch and Rittberger 2006: 28). Second, the increasing interface between public and private actors in societal steering is blurring quite settled divisions of responsibility (Stoker 1998: 17). Thus for states, the capacity to 'steer' policy has diminished as the locus of decision-making has diffused outwards to multiple levels and actors within a much more densely populated system of governance (Pierre and Peters 2000: 83–91). Third, many governance theorists locate government and governance on a continuum of state-society interactions. At one end, the centralized 'strong state' (Pierre and Peters 2000: 25) can be contrasted with the self-governing network which involves 'not just influencing government policy but taking over the business of government' (Stoker 1998: 23). Finally, governance is very often associated with the spread of non-regulatory policy instruments, or 'soft policy' as it is increasingly known (Kohler-Koch and Rittberger 2006).

MLG builds upon these arguments by extending the empirical focus to the interaction *between* the different spatial levels. Like governance, the term MLG is now 'common currency amongst scholars' and a popular means of characterizing the EU (Hooghe and Marks 2003: 234). Implicit within Hooghe and Mark's notion of MLG is the 'unraveling of central state control' (ibid.: 234), partly reflecting the broader shift from government to governance noted above. On their view, member states must share EU policy-making with supranational actors, most notably the EU Commission, sub-national governments and, increasingly, interest groups (Hooghe and Marks 2001).

Multi-level environmental governance in the EU

Many scholars have used the term MLG to make sense of the EU's environmental policy-making capabilities. Weale et al. (2000: 15), for example, maintain that the EU provides a 'multi-level and horizontally complex system of environmental governance that ... [is] none the less evolving and incomplete'. But the EU's environmental policy was not—at least initially—a system of MLG. On the contrary, it evolved via a number of different stages from its initial inception in 1973, when it looked much more like an international environmental regime than a system of governance (Jordan 2005; Hildebrand 2005).

Its subsequent development has been intricately bound up with the growth of the EU from initial beginnings as a trade-based international organization. Environmental concerns were not explicitly recognized in its founding treaty— the 1957 Treaty of Rome—although several 'incidental measures' related to agriculture, energy and trade were introduced (Hildebrand 2005: 22). Responding to an increased environmental consciousness in Europe and at the behest of member state governments, the EU formulated a programme of action in 1973. Thereafter, environmental policy grew slowly, with measures introduced in an 'ad hoc and incremental manner' throughout the late 1970s (Jordan 2005: 4). Expansion occurred despite the absence of an unambiguous legal basis in the treaty, something which some member states consistently complained about at the time (Haigh 2004: 2.1–1).

By the 1980s, the pace of legislative activity had quickened, propelled by demands from 'greener' states (ibid.) and the EU's need to harmonize laws ahead of completion of the single market. EU environmental policy was finally given a formal legal basis through its inclusion in the Single European Act 1986. Although this did not entirely please the member states (see below), the scope of EU environmental policy continued to grow throughout the 1990s. By the early 2000s, EU tasks had evolved to include measures addressing such diverse topics as water and air quality, waste management, chemicals, hazardous substances, Genetically Modified Organisms, radiological protection, wildlife protection, animal welfare, public access to information and noise (Haigh 2004; Jordan 2005). In the 2000s, EU policy-makers have sought to adopt a more strategic approach which brings together all these activities, perhaps best represented by the appearance of the EU's sustainable development strategy (2001) and (in 2005) the seven Thematic Strategies on cross-sectoral sustainability issues such as the marine environment and urban regeneration.

Mapping the 'contours' of this evolving governance system is by no means straightforward. Its architecture is characterized by multiple interacting levels, primarily the national (member state) and the EU. In this arrangement, states have opted to share their policy-making powers with other actors and institutions, most notably the EU Commission (responsible for proposing policy), the directly elected and increasingly influential European Parliament, and interest groups such as non-governmental organizations (NGOs), regional governments and industry representatives. Further EU-level control is exercised through the European Court of Justice, which has powers to force recalcitrant states to implement EU policies (see, for example, Krämer 2002). This system of policy-making today is therefore marked by horizontal and vertical complexity. But while most policy-making activity occurs at the EU level, member state governments remain important through their veto-wielding role in the Council. They are also responsible for implementing policy.

In recent years, multi-level environmental governance has become characterized by growing 'flexibility' (the sharing of policy-making with lower levels) and 'inclusivity' (the sharing of policy-making with multiple interests). In terms of output—the 'hardware' of environmental governance (Jordan 2005: 5)—these actors have helped shape a burgeoning *acquis communautaire* of environmental rules (Haigh 2004) which now largely delineate the spaces in which political contestation occurs in the member states and at EU level.

Given the multi-levelled nature of the EU, at first sight MLG would appear to be uniquely well positioned to offer significant insights into the functioning of this system. Crucially, unlike the previous generation of EU scholars who were mostly preoccupied with understanding what the EU was (an international system? A federation?) and how it was formed, those adopting an MLG perspective are keen to know how it impacts on politics, within and, most importantly, *across* the increasingly blurred 'divide' between the international and the national levels of policy-making. However, while MLG has a number of obvious strengths, it also has some weaknesses (compare Jordan 2001 and Fairbrass and Jordan 2004 with George 2004). In the next section, we take the concept of MLG and see what light it sheds on the three sites of politics noted above.

POLICY FORMULATION: WHICH LEVEL DOES WHAT?

By implication, the term MLG implies that policy-making is somehow shared between actors, both state and non-state. While these patterns are obvious in EU environmental governance, what is less certain is who actually wields power within this relationship. In other words, policy-making may not be genuinely shared, but rather still largely determined by national governments. Assessing 'which level does what?' should therefore shed more light on this issue.

Authors such as Moravcsik (1998) have concurred with a government-centric view, positing a competing theoretical analysis, liberal intergovernmentalism (LI), which privileges national state preferences as the primal source of policy. In this explanation, national governments act as the conduit through which domestic preferences are aggregated and then bargained at the EU level in an iterative 'two-level game' (Rosamond 2000: 136). On this view:

'states direct the course of integration ... are rational, self-interested actors ... keep the gate between national and international politics ... enjoy little flexibility in making concessions ...' (Fairbrass and Jordan 2004: 153).

Importantly, states are strengthened, not weakened, by such interactions (ibid.). EU policy-making, therefore, is not so much multi-level *governance* as multi-level *government* (Jordan 2001).

One way of assessing MLG and LI arguments is to focus on task allocation, which is essentially *the* central question in EU integration studies (Schimmel-fennig and Rittberger 2006). Although definitions of tasks vary, in this context it refers to the ability of an actor to formulate standards. As such, it provides a relative indicator of their policy-making power. By its very nature, integration involves a (re)allocation of tasks from member states to the EU. Many oft-cited studies have consequently sought to use task allocation as an indicator of the deepening and widening scope of the integration process (for example, Lindberg and Scheingold 1970). Integration theories have then been developed to explain this phenomenon across sectors and levels of governance (Schimmelfennig and Rittberger 2006: 76). Examination of task allocation can therefore tell us much about the claims made by advocates of an MLG perspective. For instance, if MLG theorists are correct, tasks should be allocated to multiple actors, whereas LI arguments would be supported if national government interests still dominate this process.

Before 2000, historical evidence of environmental policy-making would imply that state influence over task allocation did decline: a feature visible in the so-called subsidiarity debate (Benson and Jordan 2008b). The rapid expansion of EU tasks in the 1970s and 1980s did not occur without member state resistance. Resentment built in national governments over the costs, democratic legitimacy and effectiveness of EU task allocation, exemplified by the acrimonious conflict over the Large Combustion Plant Directive 1988 (Böhmer-Christiansen and Skea 1991).

States reacted by forcing the EU to insert what became known as the subsidiarity principle into the Single European Act 1987 to shape all future environmental policy-making (Golub 1996; 690). Subsidiarity is an organiza-tional principle derived from federal theory which states that within multiple political orders, tasks should be allocated to the lowest appropriate level unless they are more effective to address at a higher level (Follesdal 1998). The expectation by some member states at the time was that subsidiarity could be invoked to provide a brake on EU task allocation, although the reality proved somewhat different. Pressure to complete the single market allowed the EU to expand environmental policy-making into yet more areas. Tensions finally erupted in negotiations over the Treaty on European Union (TEU) in 1991. The treaty would have involved significant further transfers of tasks from member states, leading member state governments, influenced by increasingly 'Eurosceptic' publics, to force the extension of subsidiarity to all policy-making in its provisions. Subsidiarity was then subsequently used by the United Kingdom, France and Germany to produce hit-lists demanding the 'repatriation' of more controversial environmental tasks in 1992 (Jordan 2000: 1,307). Yet, the subsidiarity debate did not culminate in a wholesale dismantling of the environmental *acquis communautaire*. Very few measures were 'repatriated' (Benson and Jordan 2008b), and some new ones were introduced (e.g. the Zoos Directive), despite considerable countersubsidiarity arguments (Benson 2007). Others like the Drinking and Bathing Water Directives were amended and otherwise strengthened (Jordan 2000). Moreover, EU task allocation continued to expand inexorably throughout the 1990s. It would appear that, *contra* Moravcsik (1998), task allocation became less subject to governmental control and more multi-actored.

However, events since 2000 suggest member states have started to reassert their hold over environmental task allocation. Growing anxieties over task allocation prompted by broader globalization concerns led national governments meeting at the Lisbon European Council (2000) to ask the Commission to revise its regulatory strategy (CEC 2001). The Commission responded by publishing an action plan on subsidiarity and decision-making. In 2003, an inter-institutional agreement entitled 'Better Regulation' was concluded, based on the Commission's action plan. The Better Regulation initiative, launched in 2005, introduced several mechanisms for assessing (i.e. controlling) task allocation, for example impact assessments of proposed policies. In the environmental policy sector, the effects have been especially marked. Directorate-General (DG) Environment has increasingly been forced to defend its policy-making by proving the cost-effectiveness of potential measures in impact assessments. In addition, the language of Better Regulation has been used by powerful industrial lobbies, backed by governments, to question the upward flow of tasks. A deregulatory agenda for environmental policy has permeated the Barroso Commission, spearheaded by the DGs mostly closely associated with industry (*ENDS Report* 2005). All in all, the ability of EU actors to reallocate tasks seems increasingly circumscribed by national government preferences.

In short, the empirical support for the existence of MLG is mixed. On one hand it is undeniable that most environmental tasks have been reallocated upwards to the EU level. Since 1973, the EU has assembled a comprehensive *acquis communautaire* that has in turn significantly determined how member states manage environmental problems. As such, member states have had to learn to share tasks with multiple actors, both state and non-state, in the resolution of policy. In these terms, the multi-level governance arguments of Hooghe and

Marks et al. carry considerable conviction. Similarly, much lobbying and networking activity transcends the national-supranational divide (Fairbrass and Jordan 2004). On the other hand, we should not entirely overlook the controlling influence of national governments. While clearly no longer the 'gate-keepers' through which all decisions flow, recent events demonstrate that task allocation may not have entirely escaped their grasp.

POLICY INTERACTION: THE EUROPEANIZATION OF ENVIRONMENTAL GOVERNANCE?

The growing interest in the domestic impacts of European integration has made 'Europeanization' something of a 'hot topic' in EU scholarship (Bulmer and Radaelli 2004: 632). Judging by the recent proliferation of books and articles bearing the word 'Europeanization' (for empirical evidence, see Featherstone (2003) and Mair (2004)), it is undoubtedly one of *the* key themes in EU scholarship. However, as is often the case with 'hot topics' the meaning of Europeanization remains contested. After almost a decade of detailed scrutiny, there is, for example, still no agreed 'theory' of Europeanization (Radaelli 2003; Bulmer and Lequesne 2005: 11).

While the field of research around the term Europeanization has still not reached a mature stage, it has none the less provided important and puzzling insights into the changing nature of MLG. The oldest and the most widely adopted definition of Europeanization emphasizes the top-down impact of the EU on its member states: 'the process of influence deriving from European decisions and impacting member states' policies and political and administrative structures' (Héritier et al. 2001: 3) (for a review of alternative conceptions, see Bache and Jordan 2006, Chapter 2). Bulmer and Radaelli (2004: 4) suggested that Europeanization consists of:

'processes of a) construction b) diffusion and c) institutionalization of formal and informal rules, procedures, policy paradigms, styles, 'ways of doing things' and shared beliefs and norms which are first defined and consolidated in the EU policy process and then incorporated in the logic of domestic (national and subnational) discourse, political structures and public policies.'

In other words, the process of domestic change is said to originate in the EU and travels down through successively lower levels to the member states.

To what extent has the EU Europeanized national environmental policy? Jordan and Liefferink's (2004) comparative study examined the Europeanization of national policy. To give sharpness to their analysis, they looked at the impact on the structures, style and the content of national policies. Following Hall (1993), they defined the *content* of national policy as the overall goals that guide policy, the instruments or techniques by which these policy goals are attained, and the calibration of specific policy instruments, e.g. the level of an emission standard or tax, etc. They defined *structures* as formal bureaucratic organizations through to more informal phenomena such as codes and conventions. Finally, their definition of policy *style* drew upon what Richardson (1982: 2) originally described as society's 'standard operating procedures for making and implementing policies'. By that, he meant: (1) the dominant approach to problem solving, ranging from anticipatory/active to reactive, and (2) the government's relationship to other actors, characterized by their inclination either to reach consensus with organized groups or to impose decisions.

The emerging literature on Europeanization (Knill 2001; Boerzel 2005; Jordan 2002) has revealed that the EU's impact on these three dimensions of national policy is not at all uniform. Rather, it has affected the content of national policy much more deeply than national policy structures and policy style. Thus, the effects of Europeanization are more apparent in relation to policy goals and the calibration of tools, than policy paradigms and the introduction of new policy tools. The impact on structures has been less dramatic, incremental and mostly path-dependent. Thus, the really big 'machinery of government' changes—e.g. the creation of new ministries or the merging of existing ones—have been triggered by domestic and mostly 'non-environmental' political demands. Finally, policy style appears not to have changed significantly, although it is difficult to disentangle the 'EU effect' from other causal factors (Jordan and Liefferink 2004).

The first and most obvious explanation is that this pattern reflects the EU's primary *modus operandi*—that of an active disseminator of policy content (and especially goals and targets), not policy structures or a fresh policy style. By contrast, the EU has very little power to dictate the structure or the functioning of national public administrations, or directly influence the policy style of a country. Recall that directives (the main instrument of EU environmental policy) are mainly output-orientated—they specify the ends to be achieved, not the means of doing so. Moreover, they often contain myriad 'escape devices' (Rehbinder and Stewart 1985: 255) (e.g. the designation of improvement zones, or designated protection sites), that states can and do use to 'fine-tune' the EU's requirements to their national circumstances. So, one way or another, EU environmental policy is highly *differentiated* well before it is implemented (differentially!) at the national level (Krämer 2000).

Second, states do not passively 'take' Europeanization: they are all (to varying degrees) engaged in a 'regulatory competition' to set (or 'domesticate') the European 'rules of the game' by uploading their national policy models to Brussels. In general, the countries that have been most deeply engaged in uploading policies to Brussels are also those that are the least Europeanized. Simplifying greatly, the well-known 'pioneers' (i.e. Germany, the Netherlands and Sweden) have had to adapt the least, whereas some aspects of policy in Spain and Ireland have been completely transformed by EU membership. These two groups have been labelled as policy 'shapers' and policy 'takers' (Jordan and Liefferink 2004).

POLICY INSTRUMENTATION: MULTI-LEVEL STEERING OR A 'REGULATORY STATE'?

Another important aspect of MLG concerns the political process of selecting suitable policy instruments. We have already noted the strong tendency in the literature to associate government with (traditional forms of) regulation, whereas one of the most salient manifestations of governance is said to be the appearance of 'new' policy instruments (Zito et al. 2003). Heywood (2000: 19), for example, regards the 'ability to 'make law (legislation), implement law (execution) and interpret law (adjudication)' as the 'core functions' of government. Governance, by contrast, is characterized by a growing use of non-regulatory policy instruments such as New Environmental Policy Instruments (NEPIs). These are proposed, designed and implemented by non-state actors, sometimes working alongside state actors, but sometimes also independently. For Stoker (1998: 17), the very '*essence* of governance is its focus on governing mechanisms *which do not rest on recourse to the authority and sanctions of government*' (emphasis added).

Since the 1990s, the EU has introduced several NEPIs while promoting their broader use in member states. It could be expected, therefore, that if there is any discernible shift towards governance, use of NEPIs will have become more widespread. In their empirically informed analysis of governance patterns in the EU, Jordan et al. (2003) therefore concentrate on measuring the adoption of four different types of NEPIs: namely market-based instruments (MBIs), eco-labels, environmental management systems (EMS) and voluntary agreements (VAs). The popularity of NEPIs has grown rapidly, particularly in the last decade (Jordan et al. 2005), with environmental taxes, VAs and eco-labels proving especially popular. However, they discovered that no single type of NEPI is overwhelmingly popular in all states. In fact, some types of MBI (e.g. tradable permits) have only recently been deployed, while some 'old' policy instruments (e.g. subsidies) remain (although they are very much discredited). In some countries, the adoption of NEPIs has been stunningly fast, whereas in others, they are being adopted much less quickly.

If the adoption of NEPIs is employed as a simple measure of governance, then clearly there has been no wholesale and spatially uniform shift from government to governance across our nine jurisdictions. The overall pattern is much more highly differentiated, with regulation remaining dominant in just about all member states (Jordan et al. 2003). One obvious explanation for this is that regulation is often very hard to eliminate once it is in place. First, environmental groups in particular believe that regulation morally penalizes polluters in a way that tradable permits and voluntary agreements do not. Second, regulation (and government more generally) often provide(s) an important, but very often neglected, support function as regards NEPI use (i.e. there is significant *co-existence*). Among other things, it often provides formal authority to the agency tasked with designing and implementing a NEPI, and establishes the rules governing its operation. For example, the EU's Eco-management and Audit (EMAS) system, while formally remaining voluntary, requires member states to take various actions, such as creating an accreditation system and a certification body. A third explanation is that many environmental policy-makers are, in Herbert Simon's apt phrase, as likely to be satisfiers as utility maximizers. That is to say, while they recognize that regulation is imperfect, many still regard the case for adopting certain types of NEPI as lacking validity. In Europe, tradable permitting is still largely unproven as a general policy instrument, although this might well change now that the EU's emissions trading scheme is up and running.

The EU's role in facilitating and/or retarding the shift to NEPIs is quite intriguing. Regulation remains the mainstay of EU environmental policy (see Haigh 2004) in spite of substantial (but differential) NEPI adoption at the national level. One obvious reason is that in the case of NEPIs, the EU is ultimately dependent on state and industry co-operation for successful implementation. As a result, VAs are often technically complex to negotiate across borders, especially when well-established large industry associations are absent. Moreover, unlike more fully federated political systems, the EU has only a limited capacity to fund such initiatives since it has limited redistribution capabilities, while its powers to set taxes in member states are similarly constrained. We have already noted that regulation remains the mainstay of EU environmental policy in spite of substantial (but differential) NEPI adoption at the national level and the European Commission's White Paper on European Governance. Moreover, until now, a minority of states (initially the United Kingdom, but now also Ireland and Spain) have managed to block the Commission's ability to innovate with environmental taxation which, unlike most aspects of EU environmental policy, still falls under the unanimity rule. In fact, the EU's reliance on regulation is so deeply rooted that it has to implement many of its NEPIs (such as the eco-labelling, emissions trading and EMAS schemes) using different forms of regulation (i.e. fusion). So, far from being a case of 'new governance' (Hix 1998), the EU's experience with NEPIs underlines just how strongly constrained it is by member state (i.e. government) preferences.

CONCLUSIONS

MLG provides a useful way to begin thinking about policy-making in an era of globalization. Although increasingly used in a number of contexts, it is still most strongly associated with empirical and theoretical work on the EU, where the distinction between the national and the supranational is now arguably much more blurred than in any other international organizational system. In this essay, we have explored some of the most pertinent sites of political contestation in the EU's system of MLG, namely those relating to: *policy formulation*; *policy interaction*; and *policy instrumentation*.

In seeking to understand these sites our analysis has raised several counter-intuitive findings. First, in relation to *policy formulation*, while the EU is undoubtedly multi-levelled, the issue of scale is far from resolved, even though the EU is now by far and away *the* dominant level of policy-making. But although policy-making has become steadily more multi-actored and multi-levelled, task allocation has *not* entirely escaped the controlling power of member state governments.

Second, in relation to *policy interaction*, the shift towards MLG has significantly affected national policies. But multi-level governance should not be equated with uniform governance: despite 30 years of integration, national environmental policies have not converged upon a single, European 'model'. Having said that, even those states with long-established and environmentally ambitious policies have been forced to adjust their domestic practices so they align more closely with EU policies.

Finally, as regards *policy instrumentation*, although NEPIs have grown in popularity in EU states, evidence suggests their presence remains marginal compared to traditional forms of regulation. Furthermore, patterns of NEPI uptake are uneven, which suggests that there has not been a widespread or uniform move towards 'governance'. The EU's ability to promote such a paradigmatic shift at national levels has also been constrained: it may well exhibit several features of governance, but it remains *au fond* a 'regulatory state' (Majone 1996). It does not, for example, have hugely significant funds of its own to disburse, being mostly reliant on national governments for its funding. Moreover, member state preferences have significantly curtailed its ability to utilize NEPIs; it is therefore unwise to describe the EU as an entirely novel or unconstrained example of 'new governance' (Hix 1998).

However, despite these inconsistencies MLG does provide a potentially useful theoretical and analytical lens for viewing environmental politics, not just in the EU but in many comparable contexts. One interesting feature about the recent development of environmental policy-making globally is that it has continued to become more multi-levelled and multi-actored, as the scale of policy problems, allied to growing political, social and economic interconnectivity, has necessitated greater co-operation and co-ordination. Just as in the EU, debates over policy formulation and task allocation, processes of policy interaction and convergence, and policy instrument choice look set to remain an integral part of the design, structure and functioning of other multi-level systems. While the EU remains a very unique political system, using the concept of MLG to understand how such interactions ultimately define the scope for political actions (and hence political solutions) to problems in comparable systems, could then form the basis of future comparative research activities.

Acknowledgements

The authors would like to thank the editor and an anonymous referee for their very helpful comments on an earlier version of this essay. They would also like to thank the ESRC for supporting David Benson's Post Doctoral Fellowship (PTA-026-27-1440) and Ph.D. Studentship (PTA-042-2002-00005). Additional support was gratefully received from the ESRC-sponsored Programme on Environmental Decision Making (PEDM) at CSERGE, University of East Anglia, 2001–07 (M535255117).

Bibliography

Bache, I., and A. J. Jordan. (Eds). 2006. *The Europeanization of British Politics*, Palgrave: Basingstoke.

Benson, D. 2007. *Task allocation in the European Union: testing the explanatory value of federal theories*. Unpublished Ph.D. thesis. Norwich: UEA.

Benson, D., and A. Jordan. 2008a. Understanding Task Allocation in the European Union: testing the value of a federal perspective. *Journal of European Public Policy* (in press).

Benson, D. 2008b. 'Subsidiarity as a 'scaling device' in environmental governance: the case of the European Union'. In J. Gupta and D. Huitema (Eds), *Can theoretical insights on scale improve environmental governance?*, Boston MA: The MIT Press.

Boerzel T. 2005. Pace setting, foot dragging and fence sitting: member state responses to Europeanization. In A. Jordan (Ed.), *Environmental Policy in the European Union: Actors, Institutions & Processes* (2nd edn). London: Earthscan.

Böhmer-Christiansen, S., and J. Skea. 1991. *Acid Politics: environmental and energy policies in Britain and Germany*. London: Bellhaven Press.

Bulmer, S., and C. Lesquesne. 2005. 'The EU and its Member States: An Overview'. In S. Bulmer and C. Lesquesne (Eds), *The Member States of the European Union*, Oxford: Oxford University Press.

Bulmer, S., and C. Radaelli. 2004. 'The Europeanisation of National Policy?'. *Europeanisation Online Papers*, No. 1/2004.

Commission of the European Communities (CEC). 2001. *European Governance: A White Paper*. Brussels: CEC.

ENDS Report (2005). 'When 'better regulation' means less regulation'. *ENDS Report*, 371: 27–29.

Fairbrass, J., and A. J. Jordan. 2004. 'Multi-level Governance and Environmental Policy'. In I. Bache and M. Flinders (Eds), *Multi-level Governance*. Oxford: Oxford University Press.

Featherstone, K. 2003. 'Introduction: In the Name of "Europe"'. In K. Featherstone and C. Radaelli (2003), *The Politics of Europeanization*, Oxford: Oxford University Press.

Follesdal, A. 1998. 'Survey Article: Subsidiarity'. *The Journal of Political Philosophy,* 6 (2): 190–218.

George, S. 2004. 'Multi-level Governance and the European Union'. In I. Bache and M. Flinders (Eds), *Multi-level Governance*. Oxford: Oxford University Press.

Golub, J. 1996. 'Sovereignty and Subsidiarity in EU Environmental Policy'. *Political Studies,* XLIV: 686–703.

Haigh, N. 2004. *Manual of Environmental Policy: The EU and Britain*. London: Maney Publishing and Institute for European Environmental Policy (IEEP).

Hall, P. 1993. 'Policy paradigms, social learning and the state', *Comparative Politics*, 25 (3): 275–296.

Héritier, A., D. Kerwer, C. Knill, D. Lehmkuhl, M. Teutsch and A. C. Douillet. 2001. *Differential Europe. The European Union Impact on National Policy-making*. Lanham, MD: Rowman and Littlefield.

Heywood, A. 2000. *Key Concepts in Politics*. Basingstoke: Palgrave.

Hildebrand, P. M. 2005. 'The European Community's Environmental Policy, 1957 to '1992': From Incidental Measures to an International Regime?' In A. J. Jordan (Ed.), *Environmental Policy in the European Union: Actors, Institutions & Processes* (2nd edn). London: Earthscan.

Hirst, P. 2000. 'Democracy and Governance'. In J. Pierre (Ed.), *Debating Governance*. Oxford: Oxford University Press.

Hix, S. 1998. 'The Study of the EU II: The 'New' Governance Agenda and Its Rival'. *Journal of European Public Policy*, 5 (1), 38–65.

Hooghe, L., and G. Marks. 2001. *Multi-level Governance and European Integration*. Lanham, MD: Rowman and Littlefield.

Hooghe, L., and G. Marks. 2003. 'Unraveling the Central State, but How? Types of Multi-Level Governance'. *American Political Science Review*, 97 (2): 233–243.

Jordan, A. 1999. 'European Community water standards: Locked in or watered down?' *Journal of Common Market Studies*, 37 (1): 13–37.

Jordan, A. 2000. 'The politics of multilevel environmental governance: subsidiarity and environmental policy in the European Union'. *Environment and Planning A*, 32: 1307–1324.

Jordan, A. 2001. 'The European Union: An Evolving System of Multi-level Governance or Government?' *Policy and Politics*, 29 (2): 193–208.

Jordan, A. (Ed.). 2002. *Environmental Policy in the European Union: Actors, Institutions & Processes*. London: Earthscan.

Jordan, A. 2005. 'Introduction'. In A. Jordan (Ed.), *Environmental Policy in the European Union: Actors, Institutions & Processes* (2nd edn). London: Earthscan.

Jordan, A. 2006. 'Environmental Policy'. In I. Bache and A. Jordan (Eds), *The Europeanization of British Politics*. Basingstoke: Palgrave.

Jordan, A., and D. Liefferink (Eds). 2004. *Environmental Policy in Europe: The Europeanization of National Environmental Policy*. London: Routledge.

Jordan, A., R. Wurzel, and A. Zito (Eds). 2003. *New Instruments of Environmental Governance*. London: Frank Cass.

Jordan, A., R. Wurzel and A. Zito 2005. 'Environmental governance … or government? The international politics of environmental instruments'. In P. Dauvergne (Ed.), *Handbook of Global Environmental Politics*. Cheltenham: Edward Elgar.

van Kersbergen, K., and F. van Waarden. 2004. '"Governance" As A Bridge Between Disciplines.' *European Journal of Political Research*, 43, 143–171.

Knill, C. 2001. *The Europeanization of National Administrations.* Cambridge: Cambridge University Press.

Kohler-Koch, B., and B. Rittberger. 2006. 'Review Article: The 'Governance Turn' in EU Studies'. *The Journal of Common Market Studies*, 44, Annual Review: 27–49.

Kooiman, J. 2003. *Governing as Governance.* London: Sage.

Krämer, L. 2000. 'Differentiation in EU environmental policy', *European Environmental Law Review*, 9 (5): 133–140.

Krämer, L. 2002. *Casebook on EU Environmental Law.* Oxford: Hart Publishing.

Lindberg, L. N., and S. Scheingold. 1970. *Europe's Would-be Polity.* Englewood Cliffs: Prentice-Hall.

Mair, P. 2004. 'The Europeanization Dimension'. *Journal of European Public Policy*, 11 (2): 337–348.

Majone, G. 1996. 'A European Regulatory State?' In J Richardson (Ed.), *European Union: Power and Policy-making.* London: Routledge.

Moravcsik, A. 1998. *The Choice for Europe: Social Purpose and State Power from Messina to Maastricht.* London, UCL Press.

Pierre, J., and B. G. Peters. 2000. *Governance, Politics and the State.* Basingstoke: Macmillan.

Radaelli, C. 2003. 'The Europeanization of public policy', in K. Featherstone and C. Radaelli (Eds), *The Politics of Europeanization.* Oxford: Oxford University Press.

Rehbinder, E., and R. Stewart. 1985. *Integration Through Law, Volume II: Environmental Protection Policy.* Berlin: Walter de Gruyter.

Rhodes, R. A. W. 1996. 'The New Governance: Governing without Governance', *Political Studies*, 44, 652–667.

Richardson, J. (Ed.). 1982. *Policy Styles in Western Europe.* London: George Allen and Unwin.

Rosamond, B. 2000. *Theories of European Integration.* Basingstoke: Palgrave.

Schimmelfennig, F., and B. Rittberger. 2006. 'Theories of European Integration'. In J. Richardson (Ed.), *European Union: Power and policy-making* (3rd edn). London: Routledge.

Stoker, G. 1998. 'Governance as theory'. *International Social Science Journal*, 155: 17–28.

Weale, A., G. Pridham, M. Cini, D. Konstadakopulos, M. Porter, and B. Flynn. 2000 *Environmental Governance in Europe.* Oxford: Oxford University Press.

Zito, A., C. Radaelli and A. Jordan (Eds). 2003. 'Introduction to the Symposium on 'New' Policy Instruments in the European Union', *Public Administration*, 81 (3), 509–511.

Mass Media and Environmental Politics

MAXWELL T. BOYKOFF

INTRODUCTION

While organized studies of the art of communications—called rhetoric—began as early as ancient Greece and Rome, it was not until the 1920s that scholars actually began to speak of such activities as 'media', like we do today (Briggs and Burke 2005). Since Greek and Roman times, through the Middle Ages and Renaissance, media representations have encompassed a wide range of activities and modes of communication. From performance art, plays and poetry to news and debate, media portrayals have drawn on narratives, arguments, allusions and reports to communicate various themes, information, issues and events. Similarly, despite the presence of books, pamphlets, and newspapers that began to circulate from the time of the invention of the Gutenberg printing press in 1450, the ways we think of 'mass media' also came much later (Briggs and Burke 2005). Media growth during the intervening centuries faced constraints from a number of competing factors, such as strong state control over the public sphere, legacies of colonialism, low literacy rates, and technological capacity challenges (Starr 2004). None the less, rapid expansion of modern media communications in the 19th and 20th centuries set the stage for the impressive deployment of information via countless channels and outlets now dubbed the 'fourth estate' in contemporary society. Nowadays, scholars and practitioners describe mass media as publishers, editors, journalists and others who constitute this communications industry, and who produce, interpret and disseminate information, largely through newspapers, magazines, television, radio and the internet.

This essay broadly explores the dynamic terrain of mass media and environmental politics. The first section provides an overview of contexts that shape such interactions through time. The second section explores multi-scale processes—both external (e.g. political economics) and internal (e.g. journalistic norms) that shape mass media production of news. The third section surveys the dissemination and interpretation of environmental news in the public sphere through two models that map links between media and policy attention/engagement. The fourth section briefly looks at such processes through the example of Hurricane Katrina in 2005. Overall, this chapter outlines the contested spaces between media and environmental politics in order to help make sense of how media representations frame truth claims, as well as how larger political contexts influence such framing processes. 'Authority' garnering attention via mass media is not simply the 'truth' translated, but signifies a leveraged and amplified voice warranting critical consideration. In other words, the essay looks at the Lorax-like role of media as a key interpreter and actor at the interface of humans and the environment.

CONTEXTUALIZE THIS! MASS MEDIA AND ENVIRONMENTAL POLITICS

Through time, mass media coverage has proven to be a key contributor—among a number of factors—that has shaped and affected ongoing interactions within environmental politics. Moreover, mass media have influenced environmental policy and public understanding. Mass media representations frame environmental issues for policy, politics and the public, and draw attention to salient actors negotiating the spaces of environmental politics. Communications unfold within a larger political context that then feeds back into ongoing media coverage and considerations. For instance, regulatory frameworks (bounding political opportunities and constraints) along with institutional pressures (influencing political and journalistic norms) shape how these interactions unfold.

Many books, essays, media reports and texts throughout the last century have considered environmental issues, thus provoking attention and movements of environmental politics. For instance, Aldo Leopold's 'Sand County Almanac' prompted many to consider environmental stewardship through his discussion of the 'land ethic' (1949). In media studies, foundational texts from the 'Chicago School' (e.g. Mead 1934), the 'Frankfurt School' (e.g. Horkheimer and Adorno 1947) and luminaries such as Paul Lazarsfeld (e.g. Lazarsfeld and Merton 1964) and Walter Lippman (e.g. Lippman 1957) shaped thinking in politics and cultural studies as they related to modern media communications during these decades as well. Intersections between mass media and environmental politics coalesced more prominently in the 1960s and 1970s, as practitioners and researchers gained more insights into connections between human activities and environmental responses. For instance, Rachel Carson's book *Silent Spring* raised public awareness of the environmental risk from pesticide exposure, and examined how chemical industry interests influenced the lack of environmental policy action (1962). Carson's analysis (focused on the disappearance of spring bird songs from fatal toxic exposure) significantly shaped investigative environmental reporting and the profession of science journalism in the following decades up to the present (Kroll 2001). Therefore, in the last 40 years, mass media representational practices have shaped perceptions of associated issues of environment, technology and risk to humans, animals and ecosystems (Weingart et al. 2000; Schoenfeld et al. 1979). For instance, Burgess and Gold assembled an edited volume examining intersections between media and culture across many environmental issues (1985) and Nelkin wrote an influential book on reasons behind the increase of media coverage of science and technology (1987). Following on from this work, Burgess put forward a foundational and conceptual work regarding the production and consumption of environmental meaning via the media, and commented on the emerging need to examine aspects of the intersections between mass media, science, environmental politics and public citizens (1990).

Since the early 1990s, a sharp increase in research has explored the influence of mass media on environmental politics (e.g. Hansen 1991; Mazur and Lee 1993; Allan et al. 2000; Szerszynski and Toogood 2000; Davies 2001; Anderson 2006). Many studies have examined specific environmental issues. Examples include:

- agricultural biotechnology and genetically modified food (e.g. Kepplinger 1995; Cook et al. 2006; Nisbet and Huge 2006)
- climate change (e.g. Trumbo 1996; Antilla 2005; Carvalho 2005; Boykoff and Rajan 2007)
- earthquakes (e.g. Dearing 1995)
- energy (e.g. Koomey et al. 2002)
- hazardous waste (e.g. Szasz 1995)
- nanotechnology (e.g. Anderson et al. 2005; Stephens 2005)
- nuclear power (e.g. Gamson and Modigliani 1989; Entman and Rojecki 1993)
- natural hazards and disasters (e.g. Liverman and Sherman 1985; Van Belle 2000); and
- stratospheric ozone depletion (e.g. Ungar 2000).

The majority of these studies have examined print media coverage, while others have sought to examine television news (e.g. Cottle 2000; Van Belle 2000; Smith 2005; Boykoff and Boykoff 2007), and radio news coverage (e.g. Harbinson et al. 2006). Additionally, most of these assessments have focused on North American (e.g. Wilson 1995), UK (e.g. Anderson 1997; Miller and Riechert 2000; Smith 2000), EU (e.g. Hijmans et al. 2003; Brossard et al. 2004) and Australian/New Zealand (e.g. Bell 1994; Henderson-Sellers 1998; McManus 2000) contexts.

MULTI-SCALE INFLUENCES SHAPING MEDIA COVERAGE OF THE ENVIRONMENT

Interactions between media and environmental politics are complex, dynamic and messy. It is clear that environmental politics shape media reporting; however, it is also true that journalism shapes ongoing politics, policy decisions and activities. A key function of mass media coverage of the environment has been to 'frame' environmental issues for policy actors and the public. Meanwhile, various actors—both individuals and collective—seek to access and utilize mass media sources in order to shape perceptions of environmental issues contingent on their perspectives and interests (Nisbet and Mooney 2007). Framing is an inherent part of cognition, employed to contextualize and organize the dynamic swirl of issues, events and occurrences. It can be defined as the ways in which elements of discourse are assembled that then privilege certain interpretations and under-standings over others (Goffman 1974). Media framing involves an inevitable series of choices to cover certain events within a larger current of dynamic activities. Through journalistic norms and values, certain events become news stories, thereby shaping public perception (Tuchman 1978; Iyengar and Kinder 1987; Gamson et al. 1992). According to Entman, 'Framing essentially involves selection and salience ...' and it 'plays a major role in the exertion of political power, and the frame in a news text is really the imprint of power—it registers the identity of actors or interest that competed to dominate the text' (Entman 1993: 52–55). Asymmetrical influences also feed back into these social relationships and further shape emergent frames of 'news', knowledge and discourse. Thus, Bennett defines framing as the process of 'choosing a broad organizing theme for selecting, emphasizing, and linking the elements of a story such as the scenes, the characters, their actions, and supporting documentation' (Bennett 2002: 42).

For example, in the area of media coverage of stem cell research, Nisbet et al. have defined frames as 'central organizing idea(s) or storyline(s) to a controversy that provides meaning to an unfolding series of events, suggesting what the controversy is about and the essence of the issue' (2003: 38–42). Focusing on media interactions in the issue of nanotechnology, Anderson et al. point out that it is often a process of 'intense negotiation between journalists and their editors ... [to] help render 'an infinity of noticeable details' into meaningful categories' (2005: 201). The emphases on 'controversy' and 'negotiation' demonstrate the intensely politicized spaces these media-politics interactions occupy in the process of framing. Such considerations thus provide a window into principles and assumptions underlying framing of representations of environmental issues and politics. According to Forsyth, examinations of particular framings provide an opportunity to question 'how, when, and by whom such terms were developed as a substitute for reality' (Forsyth 2003: 81).

While journalists have consistently viewed their role as one of information dissemination rather than education, in fact the distinction between these roles becomes blurred in practice. As media representations, by their very nature, inherently frame various environmental issues, such practices also then contribute—among a host of factors (discussed throughout this book)—to setting agendas for considerations within environmental politics.

Media professionals—such as editors and journalists—operate within an often-competitive political, economic, institutional, social and cultural landscape. Path dependence through histories of professionalized journalism, journalistic norms and values as well as power relations shape the production of news stories (Starr 2004). Therefore, the construction of meaning and discourse—negotiated in environmental politics—derives through combined structural and agential components of mass media. These processes take place simultaneously at multiple scales. Large-scale social, political and economic factors influence everyday individual journalistic decisions, such as how to focus or frame a story with limited time to press as well as a finite number of column inches.

In terms of political economics, modern multinational media organizations—dominated by developed-country interests—have continued to consolidate. Efficiency and profit drive the production of news content, and this has had a detrimental effect on training for news professionals in covering varied news 'beats' (Gans 1979; Bennett 1996). This has thus affected coverage of environmental issues in sometimes contradictory ways (Dunwoody and Peters 1992; Dunwoody 1999). For example, in terms of quantity of coverage in developing countries, Harbison et al. posit that lack of journalist training for specialized environmental reporting has decreased the number of climate change stories in these countries (Harbinson et al. 2006). Conversely, when the issue itself pushes coverage beyond specialist reporters in the science pages and into political, business and general assignment reporters, the issue can gain increased coverage and attention. Nisbet and Huge assert that in coverage of plant biotechnology, such 'spillage' has helped to explain an increase in stories on the issue (Nisbet and Huge 2006). Generally, a lack of journalist training has hampered accurate communications of environmental issues through such constraints (Anderson 1997).

These issues intersect with processes such as journalistic norms and values, to further shape news content (Jasanoff 1996). These include 'objectivity', 'fairness', and 'accuracy'. Much as storylines are fuelled within environmental politics, the mass media play an important role in shaping this discourse as an interpreter, translator and disseminator of information. In a 2007 paper, Boykoff and Boykoff examine these journalistic norms. These are personalization, dramatization, novelty, authority-order bias and balance, and they shape both what becomes news and how that news is portrayed (Boykoff and Boykoff 2007).

The tendency to personalize stories means coverage focuses on 'charismatic humanoids', struggling in the negotiated spaces of environmental politics as well as daily livelihoods. The human-interest story often privileges individual movements over collective action in the media purview (Gans 1979). The gaze is on the individual claims-makers in the arena of environmental politics, and thus deeper structural or institutional analyses are foregone in favour of media attention on individual and 'sensationalized' stories. This connects to dramatization, where coverage of dramatic events tends to downplay more comprehensive analysis of the enduring problems, in favour of covering the surface-level movements (Wilkins and Patterson 1987). These norms intersect with the journalistic attraction to novelty (Gans 1979; Wilkins and Patterson 1987; Wilkins and Patterson 1991). Commonly, journalists mention the need for a novel 'news hook' in order to translate an event into a story.

These three norms inform authority-order bias, where journalists rely on 'official sources' (Wilkins and Patterson 1990). While in some cases these authorities step in to restore order, other times they serve to increase political concern. Freudenburg discusses embedded power and leveraged legitimacy enabling privileged constructions of 'non-problematicity' in environmental issues (Freudenburg 2000). For example, in the case of agricultural biotechnology, Priest et al. examined divergent framings of risk, from short-term and concentrated to long-term and diffuse. They found that framing depended chiefly on perspective (from that of a philosopher to that of an ecologist) as well as on a firm reliance on expert community views (Priest and Gillespie 2000). Finally, these norms intersect with the journalistic norm of 'balance', an activity that often appears to fulfil pursuits of objectivity (Cunningham 2003). With balanced reporting, journalists 'present the views of legitimate spokespersons of the conflicting sides in any significant dispute, and provide both sides with roughly equal attention' (Entman 1989: 30). In coverage of complex issues, such as stem cell research, risks and nuclear power, or genetic engineering, balance can provide a 'validity check' for reporters who are on deadline and do not have time nor scientific understanding to verify the legitimacy of various truth claims about the issue (Gamson and Modigliani 1989; Dunwoody and Peters 1992). While effective in some cases of political debates over environmental alternatives for action, the employment of this norm in covering issues such as anthropogenic climate change can serve to perpetrate informational biases in news reporting (Boykoff and Boykoff 2004). Overall, adherence to these norms contributes to 'episodic framing' of news, which means framing that fails to place stories into sufficient context (Iyengar 1991). This episodic framing can then skew media coverage, thereby influencing the ongoing dynamic and contested spaces of environmental politics in the public sphere

WHAT'S NEW(S)? INTERPRETATION OF ENVIRONMENTAL NEWS IN THE PUBLIC SPHERE

Once news texts or segments are disseminated into the public sphere, these encoded messages—television/radio broadcasts, printed newspapers/magazines, and internet communications—then compete in public arenas for attention. Considerations of the increases and decreases in media attention to environmental issues have predominantly been examined through two key theoretical models: Downs's 'Issue-Attention Cycle' (1972), and Hilgartner and Bosk's 'Public Arenas Model' (1988). These models seek to organize and make sense of the 'institutional, political, and cultural factors that influence the probability of survival of competing problem formulations' (Hilgartner and Bosk 1988: 56) within the mass media as well as environmental politics, policy and practices.

Many attempts to theorize the rise and fall of media coverage and public attention to ecological issues have relied on Anthony Downs's 'Issue-Attention Cycle'. For instance, in mapping the environmental policy-making process, Roberts relied on this model to 'provide an explanation of the waxing and waning of issues within the policy environment' (Roberts 2004: 141). In terms of 'agenda-setting' through the media, Newell has leaned on this model as an 'all-embracing explanation for the nature of media coverage of global warming', despite acknowledgement that the model fails to 'accurately depict the complexity and challenging nature of the climate change problem' (Newell 2000: 86). In describing the Issue-Attention Cycle as it relates to issues in ecology, Downs reasoned that attention to environmental issues moves through five sequential stages. first is the 'pre-problem stage', when an ecological problem exists but has not yet captured public attention. Downs posits that expert communities are aware of the risks, but this has not yet been disseminated more widely. The second phase is that of 'alarmed discovery and euphoric enthusiasm', where dramatic events make the public both aware of the problem and alarmed about it. Third, there is the 'gradual-realization-of-the-cost stage' where key actors acknowledge sacrifices and costs incurred in dealing with the problem. Fourth, there is the 'gradual-decline-of-intense-public-interest stage' where, according to Downs, actors become discouraged at the prospect of appropriately dealing with the issue, and crises are normalized through suppression and in some cases boredom. Finally, fifth is the catch-all 'post-problem stage', where the formerly 'hot' issue 'moves into a prolonged limbo – a twilight realm of lesser attention or spasmodic reoccurrences of interest'. In this stage, Downs covers all possibilities when he states that the issue 'once elevated to national prominence may sporadically recapture public interest' (Downs 1972: 39–41). This cycle is argued to be 'rooted both in the nature' of the problem and in the 'way major communication media interact with the public' (Downs 1972: 42).

This 'natural history' framework is useful perhaps in considering the intrinsic qualities of the issues themselves that influence these ebbs and flows of coverage. Yet, the Downs model does not capture the contested terrain of environmental politics upon which and from which 'alarm' and 'costs' are determined and contested. For examples, it does not account for political economic drivers as well as cultural pressures or social mores (exhibited through regional or national political differences). It also does not account for the non-linear factors that shape dynamic interactions in environmental politics via the mass media (Williams 2000). Critics have also made the point that cycles may have both sped up in recent years, as well as become less apparent (Jordan and O'Riordan 2000). Moreover, cross-cultural research has found evidence that while the Downs model appears to hold in some contexts, it does not hold in others (Brossard et al. 2004). In sum, this model is left wanting in that it is too partial an explanation, as well as too linear and rigid an interpretation of the messiness of the multiple internal as well as external factors shaping environmental policy, politics and practice. In terms of media coverage influencing public attention, understanding and engagement, it does not account for how the aforementioned journalistic norms could undergird what becomes news, rather than just the issue itself. Therefore, the entrenched use of this Downs model has been detrimental in considerations of *how* these media representations are constructed, thus contributing to possible impediments to considerations in the purview of environmental politics.

However, the 'Public Arenas Model' 'stresses the 'arenas' where social problem definitions evolve, examining the effect of those arenas on both the evolution of social problems and the actors who make claims about them' (Hilgartner and Bosk 1988: 55). This approach enables examinations of both intrinsic and extrinsic factors—as well as dynamic and non-linear influences that shape media influences on environmental policy, politics and practices. This helps move analyses beyond static representations to more accurate analytical lenses for understanding current trends, strengths and weaknesses in media coverage of environmental issues. In this Public Arenas Model, there is an accounting of dynamic and competitive processes to define and frame 'environmental problems'. Moreover, this accounts for the institutional arenas where these problems compete for attention and are negotiated. In other words, there is acknowledgement of the 'attention economy' (Ungar 1992) that brackets the quantity and quality of media coverage of environmental issues at a given time.

MEDIA COVERAGE OF 'HURRICANE KATRINA'

Media studies researchers have asserted that, 'Journalists are less adept at reporting complex phenomena ... [and] have difficulty reporting stories that never culminate in obvious events' (Fedler et al. 1997: 94). One such 'obvious event' was Hurricane Katrina, which made landfall in August 2005 in the Gulf Coast of the USA. This hurricane provides an example of an event that tapped into multiple political, economic and cultural pressures, and mobilized a number of journalistic norms, thus generating a 'whirlwind' of news coverage. Emergent questions of risk, hazards and vulnerability—and media coverage of them—served to reconstitute the terrain of environmental politics in terms of what the causes were, who was responsible, and what needed to be done. Intense devastation of human livelihoods and ecosystem services, as well as the toll on human lives, was overwhelming dramatic material for television, newspaper, radio and internet coverage.

At the level of the story, duelling authority figures and personalities such as then-director of the Federal Emergency Management Agency (FEMA) Michael Brown, New Orleans Mayor Ray Nagin, and members of the Bush Administration were 'news hooks' aplenty. The unprecedented impacts of the storm and the multiple layers of political, economic, environmental, social and cultural consequences generated voluminous media coverage overall. In terms of causes, scientific uncertainty regarding links between hurricane intensity and frequency and climate change provided opportunities for news outlets to call on a range of sources to comment. Overall, this biophysical and socio-political event spurred a 'wave' of coverage exploring each of these aspects and many more. Many of these politicized issues clearly fell into the messy arena of environmental politics within which policy actions were debated and discussed. In the USA, Juliet Eilperin reported in *The Washington Post*, 'Katrina's destructiveness has given a sharp new edge to the ongoing debate over whether the United States should do more to curb greenhouse gas emissions linked to global warming' (2005: A16). Considerations of links to environmental politics were made more explicit by comments from prominent political actors. For instance, Jürgen Trittin—Minister of the Environment in Germany—commented, 'The American president has closed his eyes to the economic and human damage that natural catastrophes such as Katrina—in other words, disasters caused by a lack of climate protection measures—can visit on his country' (Bernstein 2005: D5).

The Downs Model helps to make sense of how this unprecedented and dramatic event generated so much public interest, and gave way to discussions of socio-economic inequality and risk. Moreover, the model helps to trace how initially urgent discussions of funding allocation and rescue efforts have given way to continued and protracted assistance to reduce vulnerability to future environmental events. The Public Arenas Model enables further considerations of how various political, institutional and cultural factors—as well as actor networks, or 'claims-makers'—have competed for the framing and selection (as well as deselection) of these considerations within arenas such as environmental politics. This more expanded perspective thus provides tools for a more critical approach to both internal and external factors shaping mass media representations of the hurricane event itself, as well as the relevant and ongoing contested spaces of policy action. Overall, Forsyth has stated, 'assessments of frames should not just be limited to those that are labelled as important at present, but also seek to consider alternative framings that may not currently be considered important in political debates' (Forsyth 2003: 78).

CONCLUSION

Nisbet et al. point out in research on media coverage of stem cell research that, 'the events that take place in the policy sphere and the groups that compete in the political system are not only mirrored (or covered) in the media but also shaped by the media' (2003: 38). Through time, both internal (e.g. journalistic norms) and external (e.g. political economic) factors shaping media representations have dynamically refigured the terms of ongoing interactions in the arena of environmental politics. These have then also influenced ongoing considerations as well as challenges in environmental governance and policy action (Liverman 2004).

The parameters bounding this essay survey remain partial—a short essay such as this is inherently selective, thus privileging certain themes. However, topics discussed here can be placed further into context in a number of ways, thus considering them within a wider landscape. For instance, to consider the various facets of these complex processes of media coverage and environmental politics, it could be useful to consider the 'circuits of communication' model, developed by Carvalho and Burgess. This model illustrates three 'moments' or 'circuits' through which communications pass over time (Carvalho and Burgess 2005). Media communications first originate and second disseminate into the public sphere before, third, entering the private sphere of individual engagement. Stories and reports are assembled, compete for attention, are taken up to varying degrees in our personal lives, and feed back again through ongoing interactions over time. These feedbacks shape news framing in subsequent moments, and inform ongoing environmental science, policy and practice interactions. While this essay has focused on movements in environmental politics in the public sphere, there is a rich literature (beyond this scope) that addresses facets of individual understanding and engagement with media and environmental issues. For example, Lorenzoni and Pidgeon analysed 15 years of climate-change perception polling and research and found that 'perceptions of climate change are complex, defined by varied conceptualizations of agency, responsibility and trust. Successful action is only likely to take place if individuals feel they can and should make a difference, and if it is firmly based upon the trust placed in government and institutional capabilities for adequately managing risks and delivering the means to achieve change' (Lorenzoni and Pidgeon 2006: 88).

Overall, as outlined above, the public arena of mass media and environmental politics contains particularly dynamic and contested interactions with high stakes. The process of media framing involves an inevitable series of choices to cover certain events within a larger current of dynamic activities. Resulting stories compete for attention and thus permeate ongoing interactions between science, policy, media and the public in varied ways. This essay has sought to outline commonalities and trends within emergent research of mass media and environmental politics, and by so doing, point to opportunities for future, and needed, research undertakings.

Acknowledgements

The author thanks editor Chukwumerije Okereke for his outstanding work in pulling this volume together. He also thanks Mike Goodman for his useful comments and suggestions on the essay at various stages. He also thanks the James Martin 21st Century School research group at the University of Oxford for their support, as well as J. Timmons Roberts, Monica Boycoff, Emily Boyd and Diana Liverman.

Bibliography

Allan, S., B. Adam, and C. Carter (Eds). 2000. *Environmental Risks and the Media*. New York: Routledge.

Anderson, A. 1997. *Media, Culture, and the Environment*. New Brunswick, New Jersey: Rutgers University Press.

Anderson, A. 2006. Media and Risk. In *Beyond the Risk Society*, edited by G. Mythen and S. Walklate, 114–131. Open University Press: London.

Anderson, A., S. Allan, A. Peterson and C. Wilkinson. 2005. The Framing of Nanotechnologies in the British Newspaper Press. *Science Communication* 27 (2): 200–220.

Antilla, L. 2005. Climate of Scepticism: US newspaper coverage of the science of climate change. *Global Environmental Change, Part A: Human and Policy Dimensions* 15 (4): 338–352.

Bell, A. 1994. Climate of Opinion: Public and Media Discourse on the Global Environment. *Discourse and Society* 5 (1): 33–64.

Bennett, W. L. 1996. An Introduction to Journalism Norms and Representations of Politics. *Political Communication* 13: 373–384.

Bennett, W. L. 2002. *News: The Politics of Illusion*. New York: Longman.

Bernstein, R. 2005. The View from Abroad. *New York Times*. New York. September 4: D5.

Boykoff, M. T., and J. M. Boykoff. 2004. Bias as Balance: Global Warming and the U.S. Prestige Press. *Global Environmental Change* 14 (2): 125–136.

Boykoff, M. T., and J. M. Boykoff. 2007. Climate Change and Journalistic Norms: A Case Study of U.S. Mass-Media Coverage. *Geoforum* (in press).

Boykoff, M. T., and S. R. Rajan. 2007. Signals and Noise: Mass-media coverage of climate change in the USA and the UK. *European Molecular Biology Organization Reports* 8 (3): 1–5.

Briggs, A., and P. Burke. 2005. *A social history of the media: from Gutenberg to the Internet.* Cambridge, UK: Polity Press.

Brossard, D., J. Shanahan and K. McComas. 2004. Are Issue-Cycles Culturally Constructed? A Comparison of French and American Coverage of Global Climate Change. *Mass Communication & Society* 7 (3): 359–377.

Burgess, J. 1990. The Production and Consumption of Environmental Meanings in the Mass Media: A Research Agenda for the 1990s. *Transactions of the Institute for British Geography* 15: 139–161.

Burgess, J. A., and J. R. Gold (Eds). 1985. *Geography, the Media & Popular Culture.* Croom Helm: London.

Carson, R. 1962. *Silent Spring.* New York: Houghton Mifflin Company.

Carvalho, A. 2005. Representing the politics of the greenhouse effect. *Critical Discourse Studies* 2 (1): 1–29.

Carvalho, A., and J. Burgess. 2005. Cultural circuits of climate change in UK broadsheet newspapers, 1985–2003. *Risk Analysis* 25 (6): 1,457—1,469.

Cook, G., P. T. Robbins and E. Peiri. 2006. Words of Mass Destruction: British Newspaper Coverage of the Genetically Modified Food Debate, Expert and Non-Expert Reactions. *Public Understanding of Science* 15: 5–29.

Cottle, S. 2000. TV News, Lay Voices and the Visualisation of Environmental Risks. In *Environmental Risks and the Media*, edited by S. Allan, B. Adam and C. Carter, 29–44. London: Routledge.

Cunningham, B. 2003. Re-thinking objectivity. *Columbia Journalism Review* 42 (2): 24–32.

Davies, A. R. 2001. Is the media the message? Mass media, environmental information and the public. *Journal of Environmental Policy & Planning* 3 (4): 319–323.

Dearing, J. W. 1995. Newspaper Coverage of Maverick Science: Creating Controversy through Balancing. *Public Understanding of Science* 4: 341–361.

Downs, A. 1972. Up and Down with Ecology - The Issue-Attention Cycle. *The Public Interest* 28: 38–50.

Dunwoody, S. 1999. Scientists, Journalists, and the Meaning of Uncertainty. In *Communicating Uncertainty: Media Coverage of New and Controversial Science*, edited by S. M. Friedman, S. Dunwoody and C. L. Rogers. Mahwah, NJ: Lawrence Erlbaum Associates, Inc. Publishers: 59–80.

111

Dunwoody, S., and H. P. Peters. 1992. Mass Media Coverage of Technological and Environmental Risks. *Public Understanding of Science* 1 (2): 199–230.

Eilperin, J. 2005. Severe Hurricanes Increasing, Study Finds. *Washington Post*. Washington, DC, September 16: A13.

Entman, R. 1989. *Democracy without Citizens: Media and the Decay of American Politics*. New York and Oxford: Oxford University Press.

Entman, R. M. 1993. Framing: Toward Clarification of a Fractured Paradigm. *Journal of Communication* 43 (4): 51–58.

Entman, R. M., and A. Rojecki. 1993. Freezing out the Public: Elite and Media Framing of the U.S. anti-nuclear movement. *Political Communication* 10 (2): 151–167.

Fedler, F., J. R. Bender, L. Davenport and P. Kostyu. 1997. *Reporting for the Media*. Fort Worth, TX: Harcourt Brace College Publishers.

Forsyth, T. 2003. *Critical Political Ecology: The Politics of Environmental Science*. London: Routledge.

Freudenburg, W. R. 2000. Social Construction and Social Constrictions: Toward Analyzing the Social Construction of 'The Naturalized' as well as 'The Natural'. In *Environment and Global Modernity*, edited by G. Spaargaren, A. Mol and F. Buttel, 103–119. London: Sage.

Gamson, W. A., D. Croteau, W. Hoynes and T. Sasson. 1992. Media Images and the Social Construction of Reality. *Annual Review of Sociology* 18: 373–393.

Gamson, W. A., and A. Modigliani. 1989. Media Discourse and Public Opinion on Nuclear Power: A Constructionist Approach. *American Journal of Sociology* 95 (1): 1–37.

Gans, H. 1979. *Deciding What's News*. New York: Pantheon.

Goffman, E. 1974. *Frame Analysis: An Essay on the Organization of Experience*. Cambridge: Harvard University Press.

Hansen, A. 1991. The Media and the Social Construction of the Environment. *Media, Culture and Society* 13: 443–458.

Harbinson, R., R. Mugara and A. Chawla. 2006. *Whatever the Weather: Media Attitudes to Reporting on Climate Change*. London: Panos Institute.

Henderson-Sellers, A. 1998. Climate Whispers: Media Communication about Climate Change. *Climatic Change* 40: 421–456.

Hijmans, E., A. Pleijter and F. Wester. 2003. Covering Scientific Research in Dutch Newspapers. *Science Communication* 25 (2): 153–176.

Hilgartner, S., and C. L. Bosk. 1988. The Rise and Fall of Social Problems: A Public Arenas Model. *The American Journal of Sociology* 94 (1): 53–78.

Horkheimer, M., and T. Adorno. 1947. *Dialectic of Enlightenment*. New York: Herder & Herder.

Iyengar, S. 1991. *Is Anyone Responsible?* Chicago: University of Chicago Press.

Iyengar, S., and D. R. Kinder. 1987. *News That Matters: Television and American Opinion*. Chicago and London: University of Chicago Press.

Jasanoff, S. 1996. Beyond Epistemology: Relativism and Engagement in the Politics of Science. *Social Studies of Science* 26 (2): 393–418.

Jordan, A., and T. O'Riordan. 2000. Environmental politics and policy processes. In *Environmental Science for Environmental Management*, edited by T. O'Riordan. Harlow: Prentice Hall.

Kepplinger, H. M. 1995. Individual and Institutional Impacts upon Press Coverage of Science: The Case of Nuclear Power and Genetic Engineering in Germany. In *Resistance to New Technology: Nuclear Power, Information Technology, and Biotechnology*, edited by M. Bauer, 357–377. Cambridge: Cambridge University Press.

Koomey, J. G., C. Calwell, S. Laitner, J. Thornton, R. Brown, J. Eto, C. Webber and C. Cullicott. 2002. Sorry, Wrong Number: The Use and Misuse of Numerical Facts in Analysis and Media Reporting of Energy Issues. *Annual Review of Energy and Environment* 27: 119–158.

Kroll, G. 2001. The 'Silent Springs' of Rachel Carson: Mass Media and the Origins of Modern Environmentalism. *Public Understanding of Science* 10: 403–420.

Lazarsfeld, P. F., and R. K. Merton. 1964. Mass Communication, Popular Taste and Organized Social Action. In *The Communication of Ideas*, edited by L. Bryson, 95–108. New York: Harper.

Leopold, A. 1949. *A Sand County almanac*. New York: Ballantine Books

Lippman, W. 1957. *Public Opinion*. London: Macmillan Co.

Liverman, D. M. 2004. Who Governs, at What Scale and at What Price? Geography, Environmental Governance and the Commodification of Nature. *Annals of the Association of American Geographers* 94 (4): 734–738.

Liverman, D. M., and D. R. Sherman. 1985. Natural Hazards in Novels and Films: Implications for Hazard Perception and Behaviour. In *Geography, The Media and Popular Culture*, edited by J. Burgess and J. R. Gold, 86–95. London: Croom Helm.

Lorenzoni, I., and N. F. Pidgeon. 2006. Public Views on Climate Change: European and USA Perspectives. *Climatic Change* 77 (1): 73–95.

Mazur, A., and J. Lee. 1993. Sounding the Alarm: Environmental Issues in the U.S. National News. *Social Studies of Science* 23: 681–720.

McManus, P. A. 2000. Beyond Kyoto? Media Representation of an Environmental Issue. *Australian Geographical Studies* 38 (3): 306–319.

Mead, G. H. 1934. *Mind, Self & Society from the Standpoint of a Social Behaviorist*. Chicago: University of Chicago Press.

Miller, M. M., and B. P. Riechert. 2000. Interest Group Strategies and Journalistic Norms: News Media Framing of Environmental Issues. In *Environmental Risks and the Media*, edited by S. Allan, B. Adam and C. Carter, 45–54. London: Routledge.

Nelkin, D. 1987. *Selling Science: How the Press Covers Science and Technology*. New York: W. H. Freeman.

Newell, P. 2000. *Climate for change : Non-state actors and the global politics of the greenhouse*. Cambridge: Cambridge University Press.

Nisbet, M. C., D. Brossard and A. Kroepsch. 2003. Framing Science: The Stem Cell Controversy in an Age of Press/Politics. *Press/Politics* 8 (2): 36–70.

Nisbet, M. C., and M. Huge. 2006. Attention Cycles and Frames in the Plant Biotechnology Debate: Managing Power and Participation through the Press/Policy Connection. *Press/Politics* 11 (2): 3–40.

Nisbet, M. C., and C. Mooney. 2007. Framing Science. *Science* 316 (6 April): 56.

Priest, S. H., and A. W. Gillespie. 2000. Seeds of Discontent: Expert Opinion, Mass Media Messages, and the Public Image of Agricultural Biotechology. *Science and Engineering Ethics* (64) 1: 529–539.

Roberts, J. 2004. *Environmental Policy*. London: Routledge.

Schoenfeld, A., R. Meier and R. Griffin. 1979. Constructing a Social Problem: The Press and the Environment. *Social Problems* 27 (1): 38–61.

Smith, J. (Ed). 2000. *The Daily Globe: Environmental Change, the Public and the Media*. London: Earthscan Publications Ltd.

Smith, J. 2005. Dangerous news: Media decision making about climate change risk. *Risk Analysis* 25: 1,471.

Starr, P. 2004. *The Creation of the Media: Political Origins of Modern Communications*. New York: Basic Books.

Stephens, L. F. 2005. News Narratives about Nano Science and Technology in Major US and non-US Newspapers. *Science Communication* 27 (2): 175–199.

Szasz, A. 1995. The Iconography of Hazardous Waste. In *Cultural Politics and Social Movements*, edited by M. Darnovsky, B. Epstein and R. Flacks. Philadelphia: Temple University Press.

Szerszynski, B., and M. Toogood. 2000. Global Citizenship, the Environment and the Media. In *Environmental Risks and the Media*, edited by S. Allan, B. Adam and C. Carter, 218–228. London: Routledge.

Trumbo, C. 1996. Constructing Climate Change: Claims and Frames in US News Coverage of an Environmental Issue. *Public Understanding of Science* 5: 269–283.

Tuchman, G. 1978. *Making News: A Study in the Construction of Reality*. New York: Free Press.

Ungar, S. 1992. The Rise and (Relative) Decline of Global Warming as a Social Problem. *The Sociological Quarterly* 33 (4): 483–501.

Ungar, S. 2000. Knowledge, Ignorance and the Popular Culture: Climate Change versus the Ozone Hole. *Public Understanding of Science* 9: 297–312.

Van Belle, D. A. 2000. New York Times and Network TV News Coverage of Foreign Disasters: The Significance of the Insignificant Variables. *Journalism and Mass Communication Quarterly* 77 (1): 50–70.

Weingart, P., A. Engels, and P. Pansesgrau. 2000. Risks of Communication: Discourses on Climate Change in Science, Politics, and the Mass Media. *Public Understanding of Science* 9: 261–283.

Wilkins, L., and P. Patterson. 1987. Risk analysis and the construction of news. *Journal of Communication* 37 (3): 80–92.

Wilkins, L., and P. Patterson. 1990. Risky Business: Covering slow-onset hazards as rapidly developing news. *Political Communication and Persuasion* 7: 11–23.

Wilkins, L., and P. Patterson. 1991. *Risky business: communicating issues of science, risk, and public policy.* Westport, Connecticut: Greenwood Press.

Williams, J. 2000. The Phenomenology of Global Warming: The Role of Proposed Solutions as Competitive Factors in the Public Arenas of Discourse. *Human Ecology Review* 7 (2): 63–72.

Wilson, K. M. 1995. Mass Media as Sources of Global Warming Knowledge. *Mass Communications Review* 22 (1&2): 75–89.

The Politics of Climate Change

TIM RAYNER AND CHUKWUMERIJE OKEREKE

'We have met the enemy and he is us.' Pogo

INTRODUCTION

Climate change is increasingly recognized to be the most challenging environmental problem facing humanity. Its potential impacts are both grave and broad ranging, with implications for agricultural production, water supply, forests, biodiversity, human health and international security (IPCC 2001, 2007a). Rising sea levels could threaten large coastal population centres, and result in the total disappearance of some countries, particularly small island states; crop failures could threaten millions with starvation. In escaping the consequences, millions could become environmental refugees, their movements exacerbating existing international tensions (Barnett 2001). For these reasons, climate change has been said to represent a more serious threat than global terrorism (King 2004), and to be comparable in its effects to the First and Second World Wars and the Great Depression combined (Stern 2007). As for the causes of the problem, myriad human activities are in some way implicated. This, and the profound issues of equity and justice raised, makes climate change arguably unprecedented among environmental problems in its sheer complexity.

In the early years of the 21st century, as the scientific uncertainties have lessened, the gap between the scale of current policy commitments and what ultimately appears necessary has become ever starker.[1] While the international scientific community speaks of at least 90% certainty that anthropogenic greenhouse gases—rather than natural variations—are causing the planet to warm (IPCC 2007a), and a study of the economics of climate change describes it as 'market failure on the greatest scale the world has seen' (Stern 2006: 25), delivery of even the relatively modest emission reductions embodied in the 1997 Kyoto Protocol hangs in the balance.[2] Increasing public awareness of this gap is putting political and economic élites under intense pressure to take more concerted action (World Public Opinion 2007).[3] There is a sense in which the legitimacy of the modern liberal democratic state is becoming more and more dependent on its ability to handle such problems convincingly (Dryzek et al. 2003).

In the limited space available to us, this essay offers a rapid *tour d'horizon* of some of the key issues political systems face in responding to the massive challenge of climate change. We begin by outlining the characteristics of the problem that make it such a distinctive and 'wicked' one. We then provide a brief account of the international institutional response that has emerged to deal with the threat of global climate change to date. The bulk of the essay then concentrates on highlighting the key dimensions in the development of what we term an effective governance framework for the issue, how some of the key issues have been handled to date, and the hurdles that remain to be overcome.

CLIMATE CHANGE AS A 'WICKED' PROBLEM

The term 'wicked' has been used to denote a policy problem that is especially challenging, owing, among other aspects, to its inter-relationship with other problems, the need for action by protagonists with widely different values and worldviews, and the lack of opportunity for 'trial and error' learning (Rittel and Webber 1973). Significantly, with wicked problems, scientific uncertainty and diverging values among key stakeholders mean that consensus on the nature of the problem, let alone solutions, is elusive. The concepts 'intractable policy controversy' (Rein and Schon 1995) and 'unstructured problems' (Hisschemoller and Hoppe 1996) encapsulate similar concerns that protagonists subscribing to competing problem 'frames'—each making sense in its own terms and sustained by a coherent body of evidence—will tend to talk past one another, and important public policy problems will remain unresolved.

At least six characteristics of the climate change problematic may be identified that, when taken as a whole, explain why it is distinct from previous environmental policy challenges and why easy solutions are so elusive (cf. Toth and Mwandosya 2001; Stern 2006).

(i) The causes of the problem are global
Climate change is related to the concentration of a range of greenhouse gases (GHGs) in the atmosphere.[4] This concentration is the product of emissions from all sources and from all countries, such that unilateral reductions by individuals, firms or countries have a small overall effect. Compared to other environmental pollution challenges, emissions are linked to a broader array of human activities, including transport, energy use, industrial activities, and agriculture, and have an equal effect regardless of where they originate. Moreover, most emissions come from sectors regarded as important to national security and/ or economic growth (Baumert and Kete 2002). Emissions are also influenced by a wide range of policies affecting technological innovation, economic growth and population size.

(ii) The impacts are global
They include a **significant effect on the global economy** if adequate action is not taken. Economic analyses that take into account the full ranges of both impacts and possible outcomes suggest that climate change under a 'business-as-usual' scenario would reduce welfare by an amount equivalent to a reduction in consumption per head of between 5% and 20% (Stern 2006).

117

(iii) While global in cause and effect, relative contributions to the problem and the burden of its consequences are highly asymmetrical

While the majority of the accumulated global stock of greenhouse gases has arisen from the industrialized world, it is the developing countries that have contributed least to the problem that stand to be hit the hardest by the effects of climate change; effectively, the rich impose costs on the poor. That said, countries such as China and India are now also significant emitters—albeit much lower in per caput terms than developed states.[5] Developing countries argue that while theirs are essential 'subsistence emissions', those from the developed world constitute 'luxury emissions' (Shue 1993). Who should bear the burden of emission reduction is therefore a complex question of environmental justice.

(iv) The problem is long-term

Emissions of carbon dioxide, the primary greenhouse gas, remain in the atmosphere for about 100 years. This has several implications. The first is that lines of causation between particular emissions and effects are not clear, complicating discussions over responsibility and liability. Secondly, it means that whatever measures are taken now, the world is already 'committed' to some temperature increase.[6] Complex ethical questions over 'intergenerational equity' arise because the future generations that will be most affected by climate change are not able to participate in present-day decision-making processes (Page 2006). The time-lag between emitting GHGs and the experience of their ultimate effects means that today's generation is being asked to make sacrifices, change habits, and face higher costs of carbon-intensive activities, in order principally to benefit future generations (EAC 2007).

(v) Uncertainty is pervasive

Despite advances in science, many uncertainties remain regarding the magnitude of future climate change and its consequences. One area where there is a large degree of uncertainty is the possibility of abrupt and irreversible climate change. According to Lenton et al. (2006), the possibility of abrupt climate change is as real in a scenario whereby all emissions completely cease now as it is for a business-as-usual case. This is because 'abrupt changes can be triggered many decades before they actually occur' (Lenton et al. 2006: 8). Uncertainty also permeates debates over the costs, benefits, and barriers to implementation of possible solutions. These uncertainties create the conditions in which significant and intractable 'frame-conflicts' over the nature of the problem, and whether it is being adequately dealt with, are able to persist.

(vi) The global institutions arguably necessary to address the issue are only partially formed

Environmental problems are often characterized in terms of the collective action problems inherent in the protection of public goods (see, *inter alia*, Weale 1992 and Young 1989). An environmental resource may be termed a public good where clear property rights have not been assigned, but remain in a state where one individual's consumption does not limit that of others, and where if one actor refrains from a harmful activity, others cannot be excluded from the resulting benefits. Environmental protection must overcome 'collective action problems' which arise because benefits to be gained from using a public good (e.g. using the environment as a sink) are often concentrated among a handful of producers, while costs may be spread widely. While polluters have a strong incentive and ability to act collectively to forestall policy measures that may rein in their activities, those being harmed (including future generations) find it more difficult to mobilize. Furthermore, if individuals (including citizens, firms, whole countries) cannot be excluded from benefits provided by others, each has an incentive to free-ride on efforts of others to protect a public good (Olson 1965; Weale 1992). Collective action by independent sovereign nations is particularly challenging. On climate change, since no supranational authority exists to impose coercive sanctions, co-operation requires nations to perceive sufficient benefits that they are willing to participate in international treaties or other arrangements and share a common vision of responsible behaviour.

THE EVOLUTION OF INTERNATIONAL CO-OPERATION ON CLIMATE CHANGE

Despite this formidable array of challenges, and contrary to the expectations of many, international co-operation on climate change has proved possible. In 1988, the World Meteorological Organization (WMO) and the United Nations Environment Programme (UNEP) established the Intergovernmental Panel on Climate Change (IPCC), to assess the scientific, technical and socio-economic information relevant to understanding the risks associated with human-induced climate change. The need to treat climate change as a problem in need of a concerted political response was formally recognized in the United Nations Framework Convention on Climate Change (UNFCCC), negotiated as part of the Rio 'Earth Summit' in 1992 and ratified by 189 signatories.[7] This introduced the key principles that have framed deliberations ever since: that policy responses should be designed to avoid 'dangerous' climate change, and should do so on the basis of 'common but differentiated responsibilities and respective capabilities' (Articles 2 and 3 respectively). The convention also calls on industrialized countries to 'take the lead' in protecting the climate (Article 3.1). In 1997, under the Kyoto Protocol, industrialized countries were required to reduce their emissions by 5.2% relative to 1990 levels for the period 2008–12; developing countries, in accordance with the principle of differentiated responsibilities, were left without targets for the period. Despite the decision of the US Bush Administration in 2001 not to ratify the protocol, it entered into force in 2005, having met the requirement of ratification by 55 parties to the UNFCCC, accounting for at least 55% of industrialized countries' CO_2 emissions in 1990.

In the next section, we move on to examine the nature of this emerging international governance framework in more detail, and some of the challenges ahead.

GOVERNANCE—THE KEY DIMENSIONS

The process of governance on any policy question is structured by institutions: 'the systems of rules, decision-making procedures and programs [sic] as articulated in constitutive documents', 'social practices that are based on the rules of the game but also [. . .] common discourses in terms of which to address the issues at stake, informal understandings regarding appropriate behaviour on the part of participants, and routine activities that grow up in conjunction with efforts to implement the rules' (Young 2002: 5–6). In building a 'governance framework'—the totality of governance and its institutional context—on any particular policy issue, a number of questions need to be addressed. Informed by our outline of the key characteristics of the climate change problematic, we posit the key dimensions of governance as follows:

i) Problem definition. What, precisely, is the problem; who bears responsibility for it, and should therefore have policy measures targeted at them? What priority should it be accorded in relation to other problems?

ii) Policy commitments. What kind of policy commitments should be made? What targets, if any, should be set?

iii) Levels and scales. To what extent should a harmonized, international response be pursued, or a more fragmented, differentiated approach accepted?

iv) Modes/instruments of governance. Where should the balance be struck between different types of policy instrument?

v) Implementation and enforcement. How can implementation and enforcement of policy, potentially over the long term, be assured?

Debates on these questions are pervaded by considerations of costs and benefits to various constituencies, and equity.

In the next section, we elaborate on what these dimensions entail, the difficulties and dilemmas involved in securing agreement on how to frame and act on the problem, and how they have been handled to date in the current governance framework.

BUILDING EFFECTIVE CLIMATE GOVERNANCE

Problem definition

Problems do not speak for themselves, but need to be 'constructed' or 'framed' as such by societal actors before policy responses can be contemplated. Gross air pollution, for example, is not universally regarded as problematic; at some times and in some places, it may be accepted as indicative of a healthy economy. The governance question here is whether and when particular physical or social phenomena should be regarded as 'problems' that are rightfully matters for policy intervention (Kingdon 1995; Hajer 1995), and which actors bear the responsibility to act. Resistance to accepting particular conditions as problems is likely where solutions appear costly, but developments in scientific evidence and shifts in public opinion may force a response from political actors; subtle efforts to 'reframe' available evidence may also prove effective (Schon and Rein 1995). More specifically, there is the question of which aspects of the policy 'problem' policy-makers should tackle, and with what level of priority and sense of urgency in relation to other goals. How a problem is framed determines to a large degree the kind of policy commitments that are made and the 'solutions' developed. This section highlights the central elements of this 'discursive struggle' (Hajer 1995).

Although a degree of resistance remains, the scientific consensus that recent and future temperature changes are the product of human activity, rather than natural processes, is stronger than ever.[8] However, exactly what constitutes the 'dangerous' climate change that international co-operation is meant to avoid has been contested ever since the concept was agreed as part of the UNFCCC. To a large degree it is dependent on the scale at which the problem is viewed, and on the perceptions of the stakeholders involved (Dessai et al. 2004; Schellnhuber et al. 2005). Even a relatively small 2°C warming above pre-industrial levels—at the time of writing, the lowest figure considered achievable, given radical emission cuts—may be catastrophic at some scales, for example Pacific small island states and sensitive habitats such as coral reefs. The further past 2° the planet warms, the greater the risk of irreversible, catastrophic changes, such as the collapse of the Greenland ice-sheet. The likely impacts from different degrees of warming are illustrated in Figure 1.

Figure 1. Reasons for concern about climate change impacts.

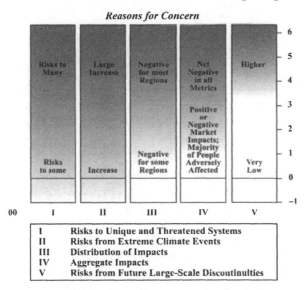

Source: IPCC (2001b) Impacts, Adaptation, and Vulnerability - Contribution of Working Group II to the IPCC Third Assessment Report, Eds J. J. McCarthy, O. F. Canziani, N. A. Leary, D. J. Dokken, K. S. White. Cambridge: Cambridge University Press.

The notion of common but differentiated responsibilities reflects the fact that although all states bear a degree of responsibility for their emissions, industrialized countries are responsible for about 63% of human-related carbon dioxide that has accumulated in the atmosphere, whereas the 80% of the world's population living in developing countries has contributed about 37% (Baumert and Kete 2002). In terms of current emission levels, Figure 2 shows that the average American emits about 10 times more carbon than the average Chinese and 20 times more than the average Indian. As noted above, the Kyoto framework recognized this asymmetry by placing emission-reduction commitments on industrialized countries only, at least during the period it covers.

Figure 2: CO$_2$ emissions per caput and population by region in 2000 (Grubb 2004).

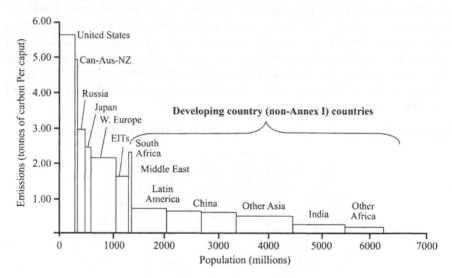

Climate change has often been framed conceptually as a market failure, caused by a failure to allocate property rights and 'internalize' external costs. In this view, the problem can in principle be dealt with, even in the context of a continually growing economy, provided appropriate measures are taken to address various market failures. The clearest statement of this view can be found in the Stern Review (Stern 2007), a report commissioned by the UK Government with a view to galvanizing international action (Jordan and Lorenzoni 2007). In accordance with the discourse of 'ecological modernization', while the structural nature of the climate crisis is increasingly recognized, and a wider range of stakeholders implicated in delivering solutions, a solution is not thought to necessitate an abandonment of the principal institutions of modernity and capitalism (Hajer 1995).

Indeed, climate policy is increasingly perceived as offering a range of economic benefits, in the form of 'new sites of accumulation' (Matthews and Patterson 2005) such as emissions trading markets, in addition to other 'co-benefits' such as health improvements (from better air quality) and improved security of energy supply (resulting from reductions in fossil fuel use). How far to emphasize such co-benefits does, however, represent a dilemma for those building a coalition in favour of radical emission cuts. Highlighting of energy security benefits, for example, can lead to solutions such as large-scale promotion of biofuels (as aggressively promoted in the later years of the Bush Administration) that potentially have adverse consequences for food security, biodiversity, even ultimately the climate (McNeely 2006). Only the more radical environmentalists seem prepared to voice the view that economic growth may be the fundamental driver of environmental degradation (Verweij et al. 2006; Sachs 1999), although calls for technological development to be at least complemented by significant changes in lifestyles in the developed world—a reduction in the distance travelled by air, for example—are more commonly heard.

Given that the perception of high costs associated with mitigation has tended to deter key actors, particularly the US Federal Government, from acknowledging climate change as a serious problem, the Stern Review's conclusion that 'the benefits of strong and early action far outweigh the economic costs of not acting'—and its favourable reception among economic and political élites (Jordan and Lorenzoni 2007)—are potentially significant. Significantly, the review also moved the debate away from the conventional discourse of cost-benefit analysis, with its insistence on costing all impacts, and discounting those occurring in the future, towards an advocacy of basing decision-making on a more nuanced appreciation of risk and uncertainty in the face of potentially catastrophic (in economists' terms, 'non-marginal') change.

Finally under this heading, the question of the relative priority accorded to mitigation (efforts to reduce emissions) and adaptation (measures to respond to changes in the climate) also deserves mention. One striking element in the framing of the problem until recently has been that detailed debate on adaptation has been subject to a kind of taboo, on the grounds that it could detract from the urgent imperative of mitigation (Pielke et al. 2007). Although recent years have seen an important shift here, underpinned by the realization that even deep cuts in emissions would fail to stop temperatures rising significantly, given the limitations of space here, we have decided to concentrate our discussion on mitigation, rather than adaptation.

Policy commitments

Given the growing consensus on the seriousness of the problem, and that it is the product of human activities, the question arises as to what high-level policy objectives should be set to address it. In this section, given limitations of space, we concentrate on current debates over emission-reduction targets. Before addressing this key question, however, possible alternative commitments and the issue of whether targets should be binding or non-binding are briefly discussed.

As far as the possible commitments that could in principle be made are concerned, suggestions have ranged from absolute caps, efficiency targets, technology development, co-ordinated carbon taxes, co-ordinated sector-specific domestic policies or a mixture thereof (see for example Egenhofer and van Schaik 2005 for a review). At Kyoto, negotiators were attracted by the relative simplicity and sensitivity to environmental integrity of absolute targets for emissions reduction. However, the disadvantages of such target-setting—principally its inflexibility with respect to inevitable differences in growth of population and/or economies—have led many, notably outside Europe, to advocate alternative ways to formulate commitments. Part of the justification offered by the Bush Administration for its notorious refusal to ratify its Kyoto targets was that a non-mandatory approach, centred on development of low-carbon technology, would be as effective and incur lower economic costs.[9] It seems clear, however, that without binding emission-reduction targets, the incentives to technological innovation and diffusion are simply inadequate (Baumert and Kete 2002; Grubb 2004).

On the question of whether commitments should be binding or non-binding, the arguments centre on the trade-off between greater certainty of delivery and the ambitiousness of the cuts agreed, a question that will be addressed in the discussion of levels and scales, below. The current global climate regime may be considered a mixture of binding and non-binding. The provisions of the UNFCCC, for example, including the initial 'aim' for industrialized countries to return their GHG emissions to 1990 levels, are widely considered non-binding (Depledge 2002)—perhaps explaining the convention's relatively unproblematic adoption. On the other hand, the subsequent Kyoto Protocol of 1997, negotiated under the UNFCCC framework, reflected an agreement that the climate change problem required a binding, target-based approach to the reduction of GHG emissions.

As already noted, under Kyoto, in accordance with the principle of differentiated responsibilities, developing countries were left without targets. In view of the criticism that has been targeted at Kyoto, it is important to note that the initial 'commitment period' was envisaged as a first stage in a permanent process of negotiations, agreements, and enforcement that would eventually include developing countries (an element that will also feature in the discussion of levels and scales below). Similarly, given criticism that the minimal size of the emission cuts it entailed did not justify the economic disruption (see for example Lomborg 2001), it is important to recognize that the rationale behind the Kyoto targets was not to achieve an 'optimal' global level of emission reduction, balancing costs and benefits of action and inaction, but to take current emission levels as the rough baseline, and to attempt in the first instance to hold that line (Helm 2003). It would be for subsequent steps to achieve more radical reductions in emissions.

On what basis, then, should the more radical reduction targets, widely regarded as necessary, be set? Much discussion is now focused on the need to agree a quantitative global stabilization of the stock of greenhouse gases, to be achieved by 2050. This could serve as a policy framework in which short-term policies to achieve emissions reductions would need to be consistent with the long-term stabilization goal. The debate is framed around balancing probabilistic estimates of temperature changes associated with particular atmospheric concentrations against the feasibility of achieving certain targets, with much discussion highlighting the desirability of aiming for a range between 450 and 550 parts per million carbon dioxide equivalent (ppm CO_2e) (Stern 2006). Although the science is fraught with uncertainty, some research suggests that even a 450ppm CO_2e target entails a more-than-even chance of exceeding a 2°C warming (Meinshausen 2005), leading many to advocate treating this as the global target (see for example *The Ecologist* 2007). The fact that by 2006 the concentration was already 430ppm CO_2e, and rising annually at more than two ppm, makes stabilization at 450ppm particularly ambitious (Stern 2007). Even the politically more realistic range of 500–550ppm necessitates radical and urgent action, including 'complete transformation of the power sector' (Stern 2006: 299). Moreover, stabilization will require *ongoing* reductions in emissions *beyond* 2050, ultimately necessitating, arguably, the complete decarbonization of every human activity beyond agriculture (ibid.).

Significantly, stabilizing greenhouse gas concentrations at a chosen level does not simply depend on how much lower annual emissions are by mid-century, but how quickly these reductions occur. The longer that global emissions continue to rise year on year, the harder (and costlier) it will be, once annual emissions begin to decline, to make annual cuts large enough and fast enough to get down to equilibrium with the carbon absorption rate in time to stabilize the final atmospheric total at a desired target level (Stern 2006).

Levels and scales of governance

Here the principal questions are, firstly, how the responsibility for delivering adequate emission reductions should be distributed spatially, and secondly, whether a co-ordinated, 'universalist' global regime or a more fragmented 'orchestra of efforts' (Sugiyama et al. 2003) is most likely to deliver effective results.

As already noted, Kyoto's exclusive focus on industrialized countries was only ever meant to be temporary, and despite continuing refusal on their part to accept absolute quantitative targets, keeping atmospheric concentrations within the kind of limits described above will inevitably require significant commitments from developing countries too. To achieve stabilization at 550ppm CO_2e in 2050, it is estimated that global annual emissions would need to be reduced by 60%–65% from business-as-usual (BAU) levels, and by more than 85% to stabilize (without overshooting) at 450ppm CO_2e (Stern 2006). Even if emissions from developed countries could be reduced to zero in 2050, the rest of the world would still need to cut emissions by 40% from BAU to stabilize at 550ppm CO_2e; for 450ppm, this rises to almost 80% (Stern 2006: 206).

One of the key dilemmas for international negotiations in the so-called 'post-2012' framework is whether to pursue agreement on a universal global regime involving binding emission-reductions—something like an expansion of the existing Kyoto framework—or to accept a greater degree of fragmentation.[10] The danger is that a universal agreement may only be possible, if at all, after protracted negotiations, and by adopting a lowest common denominator approach that is unlikely to be radical enough. The experience of Kyoto was that targets were arrived at through an essentially *ad hoc*, bottom-up negotiation process, shaped more by political power and economic might than by any objective criteria (Baumert and Kete 2002; Grubb 2004). If negotiations among developed countries were tough, including a wider set of players in building a post-Kyoto, universal regime will be far tougher. Critics have called for negotiations to be based on overarching principles or rules that, once agreed, would guide the subsequent emission-reduction efforts among nations in a more orderly fashion. The so-called Brazilian Proposal and per caput allocations are two examples of principle-based approaches, the former advocating apportioning emission-reduction requirements based on each country's relative, historic responsibility for the global temperature increase, the latter proposing distributing allowances according to the size of a country's population (Baumert and Kete 2002). The political challenges associated with adopting such principles are clearly formidable.

If pursuing a universal agreement presents dilemmas, the alternative of allowing more ambitious states (and within them, lower tiers of government), groups of states, even companies to 'go it alone' and implement a more stringent climate policy among themselves, poses risks of its own. For example, although it might allow faster agreements and greater reduction commitments per party, such an approach may be prone to international 'leakage'—the relocation of polluting industries from strictly regulated to less regulated jurisdictions. While empirical evidence on trade and location decisions in the current regulatory context is fairly reassuring on this concern (see COMETR 2007), the effect of significantly increased energy costs is uncertain. However, the competition to which key industry sectors are subject tends to come predominately from countries within regional trading blocs, which suggests that action at this level will contain the competitiveness impact.[11] The stronger the expectation of eventual global action, the less likely are firms faced with long-term decisions when investing in plant and equipment to relocate.

A further dilemma related to scale concerns the extent to which it should be permitted for emissions reductions to be achieved by investments and activities undertaken abroad, rather than domestically. The Kyoto Protocol contains provisions for the use of 'flexible mechanisms', including emissions trading (ET), Joint Implementation (JI) in projects in economies in transition and the Clean Development Mechanism (CDM) investments in developing countries.[12] While the argument here is that by allowing such flexibility, emissions-reductions will be achieved in the most cost-effective manner, critics denounce a shirking of domestic responsibility. Significantly, agreement at Kyoto was only possible by incorporating the US preference for flexible mechanisms (making subsequent US withdrawal all the more ironic) and it seems inevitable that such flexibility will continue to be a necessary sweetener for significant emission-reduction targets for developed countries in any future framework.

Different modes and instruments of governance

A further key question arising in the construction of an effective response to climate change is, which 'modes of governance' or specific instrument mix policy-makers should promote. To what extent should they seek to constrain the actions of targeted groups and individuals (as in hierarchical, regulatory modes), offer financial incentives (as in market modes), or build trust between them (through network governance) to deliver policy objectives? A further question under this heading is, to what extent policies should target small numbers of key actors, such as power plants, or attempt to change the behaviour of large numbers of citizens (Kok et al. 2002).

Historically, collective action problems have tended to be overcome by recourse to hierarchical forms of regulation, whether imposed by international level treaties, laws of national government or more local levels of jurisdiction. Such direct, 'top-down' regulation, however, is not the only possibility. For various reasons, such a response may not be regarded as feasible or desirable, especially by national-level policy-makers faced with an apparent waning of the power of the state, as control is displaced upwards to regional and international organizations, downwards to regions and devolved localities, and outwards to international corporations and other private or quasi-private bodies (Pierre and Peters 2000). Increasingly, policy-makers are attempting to exercise *meta-governance*: 'the judicious mixing of market, hierarchy, and networks to achieve the best possible outcomes from the viewpoint of those engaged' (Jessop 2003: 108). To a large degree, current responses reflect the specific nature of individual policy sectors. In the case of EU countries, while the domestic sector is dominated by regulation, the transport sector has relied heavily on voluntary and negotiated agreements. Business faces a mix of all types of instrument.

The framing of climate change as a particular kind of market failure implicates certain solutions. If carbon emissions are an unpriced externality, then arguably the first essential element of climate change policy is carbon pricing (Stern 2006). Putting an appropriate price on carbon, through taxes or emissions trading, means that the full social cost of actions is reflected in market transactions, with the result that individuals and businesses are presented with incentives to switch away from high-carbon goods and services, and to invest in low-carbon alternatives. As a first step, a range of perverse subsidies, encouraging fossil fuel use, need to be removed. This, however, is politically difficult, given that powerful constituencies benefit from them.[13]

The presence of a range of further market failures and barriers, however, tends to mean that higher carbon pricing alone is not sufficient, but needs to be complemented by, in particular, technology policy. Here, public expenditure on research and development, demonstration, and market support policies can all help to drive innovation, and motivate a response by the private sector (Kok et al. 2002). And given that opportunities for cost-effective mitigation are not always taken up, for a range of possible reasons including a lack of information, the complexity of the choices available and individual and organizational inertia, policies to promote behavioural change may be regarded as a further, critical element. Indeed, without such change, emission reductions made through improvements in eco-efficiency are liable to be eroded as consumers spend the resultant savings on additional polluting activities. Examples of such 'rebound effects' include driving further because more efficient vehicles lower total fuel costs, or buying more household appliances. Policies to encourage behavioural change may take the form of regulation (e.g. setting minimum standards for household appliances), information provision, for example through labelling, and financial incentives. Behavioural change on the part of citizens will ultimately depend on the emergence of a shared understanding of the nature of climate change and its consequences and new norms and values, fostered through education, persuasion and discussion (Stern 2007).

Kok et al. (2002) highlight the need for 'tailor-made' policies, depending on the actor, option or technology and the context. At one end of the spectrum, households have specific interests, and applying new technologies or modifying consumption patterns have specific, local requirements. The petrochemicals industry, by contrast, operates in a completely different context of a fiercely competitive global market, with effective action more likely to entail agreements across the sector, on a world-wide scale. In one sphere, the development of new technologies is a major issue, in another it is more a matter of implementing schemes and relatively minor changes in the marketplace. The fundamental issue is that specific conditions require specific management and policies to succeed. Success, according to Kok et al. (ibid.: 229), also requires a 'decisive government' that is capable of developing, implementing and maintaining such sector-specific policies in close co-operation with relevant actors in the market and civil spheres. 'Decisive government' in this case does not entail strong application of top-down command and control mechanisms, but a forward-looking government that consistently shows the direction for development. It is to the issue of consistency over time that we now turn.

Long-term implementation and enforcement issues

The need for mechanisms to guarantee compliance and sufficient resources for implementation may be considered generic policy concerns. In the case of climate policy, a particular concern is to ensure policy continuity and predictability over extended periods of time. What makes the long-term *credibility* of policy particularly important in climate change politics is that private sector actors will be required to make massive capital investments to deliver emissions reductions. Without long-term certainty that the fundamentals of the policy framework will remain in place, firms will be reluctant to make such investments. These concerns are highly germane to current emissions trading schemes, as well as policies which involve higher levels of carbon taxation. Faced with conflicting political demands, including the need to secure (re)election, governments may be tempted to renege on promises to tax carbon at certain levels or, in the case of emissions trading, on promises about the number of permits available. A credible carbon policy at national level must therefore overcome two hurdles (Helm et al. 2003). First, given that climate policy-related objectives are likely to come into conflict with longer-established policy goals, clear rules are required that allow for resolution of trade-offs. Otherwise, the danger will be that climate-related goals are consistently trumped by others. Second, and more importantly, the governments must convince firms that they will not renege on their promises once investment costs are sunk.

While apparently ambitious climate policy objectives have been adopted in various jurisdictions, the extent to which these have been prioritized over competing demands has been questionable. The situation is not helped by the absence of strict enforcement provisions at international level. Like most international treaties, the explicit consequences for non-compliance with Kyoto commitments are weak compared to domestic law; the most concrete are that failure to meet the quantified commitments in the first period automatically disqualifies a country from participating in the flexible mechanisms and that it will be penalized by deductions from allowed emissions in subsequent rounds with a 30% penalty factor (Grubb 2004: 7).

In principle, what Helm et al. call the 'time-inconsistency problem' in carbon policy can be solved through delegation to an independent agency. Such delegation reduces uncertainty and political and regulatory risk, thereby reducing the cost of capital. It reduces the possibility that governments, which are driven by the next election and other short-term political economy considerations, will set carbon policy inappropriately. An independent agency is a key element in the UK Government's pioneering Climate Change Bill, which creates a strong legal framework to underpin the United Kingdom's contribution to tackling climate change. It does this by setting out a clear and credible pathway to a statutory goal of a 60% reduction in carbon dioxide emissions through domestic and international action by 2050, with real progress by 2020. This will be based on a new system of 'carbon budgets' set at least 15 years ahead, and with progress reported annually to Parliament. A new expert 'Committee on Climate Change' will advise the Government on the best pathway to 2050 (DEFRA 2007).

CONCLUSIONS

Throughout this chapter, the term 'effective' has been used as a standard by which to judge the climate change governance framework, but never concisely defined. It is in the nature of wicked problems, however, that the effectiveness of dealing with them is always open to interpretation and contestation. Given the timescales involved, for example, it may well be that what in 2050 appears to have been an effective solution, by 2100 comes to appear inadequate. That said, for the sake of argument, in this concluding section we assume effective to mean the kind of policy response under which the containment of temperature rises to 2℃ is at least a realistic possibility.

In view of the huge collective action problem represented by climate change, the reaching of an agreement at Kyoto could be regarded as a remarkable success (Grubb 2004). However, the commitments agreed were small in relation to the emissions reductions that need to be delivered in the first half of the 21st century. In view of the obvious political sensitivities entailed in delivering radical emissions reductions, policy-makers have so far tended to pursue measures that are self-evidently cost-effective and able to deliver 'ancillary' or 'co-benefits' in addition to greenhouse gas abatement. This approach, while successful as a short-term strategy for building a political coalition behind mitigation efforts, will have to be superseded by one acknowledging the need to act even at significant immediate cost.

For such a progression to occur, a significant reframing of the climate change problem is required. Hugely costly mitigation efforts would have to be justified as an insurance policy against severe risks of catastrophic change, the precise likelihood of which is impossible to pin down.[14] The current partial coverage of international agreements would also have to give way to some kind of meaningful inclusion of the USA and key rapidly industrializing countries, principally China and India, in order that those taking firm action in more progressive states are not disadvantaged competitively. Policies and institutional frameworks would need to be credible enough to spur the new patterns of investment necessary to bring about transition to a low-carbon economy; current regimes which put a value on carbon, for example, would need to be extended. A strong argument can be made that achievement of the kind of stabilization target needed to avoid a temperature increase of more than 2–3℃ requires a universalist architecture entailing binding targets (Boeters et al. 2006). Arguably, because costs of mitigation and adaptation will fall unevenly across regions, and impacts will remain difficult to predict, agreement will depend on the emergence of a shared understanding of the nature of the problem from a global, rather than a parochial perspective; in other words, a redefinition of notions of national interest and security is required. Such a redefinition is without historical precedent, and will require considerable feats of political leadership.

Bibliography

Barnett, J. 2001. *Security and Climate Change*, Tyndall Centre Working Paper 7, Tyndall Centre for Climate Change Research, Norwich

Baumert, K. A., and N. Kete. 2002. 'Introduction: An Architecture for Climate Protection. In *Building on the Kyoto Protocol: Options for Protecting the Climate*, edited by K. A. Baumert, O. Blanchard, S. Llosa and J. Perkaus. World Resources Institute: Washington, World Resources Institute.

Boeters, S., M. den Elzen, T. Manders, P. Veenendaal and G. Verweij. 2006. *Post-2012 Climate Policy Scenarios*, MNP Report 500114006/2007, Netherlands Environmental Protection Agency, Amsterdam.

COMETR 2007. *Competitiveness Effects of Environmental Tax Reforms, Policy Brief*, National Environmental Research Institute, University of Aarhus (NERI), available at http://www2.dmu.dk/COMETR/

Depledge, J. 2002. Continuing Kyoto: Extending Absolute Emission Caps to Developing Countries. In *Building on the Kyoto Protocol: Options for Protecting the Climate*, edited by K. A. Baumert, O. Blanchard, S. Llosa and J. Perkaus. Washington, World Resources Institute.

Dryzek, J., D. Downes, C. Hunold, D. Scholsberg, and H. C Hernes. 2003. *Green States and Social Movements*. Cambridge: Cambridge University Press.

DEFRA 2007. *Draft Climate Change Bill*, Presented to Parliament by the Secretary of State for Environment Food and Rural Affairs By Command of Her Majesty, March 2007, Cm 7040.

Dessai, S., W. N. Adger, M. Hulme, J. Turnpenny, J. Kohler and R. Warren. 2004. Defining and experiencing dangerous climate change, *Climatic Change* 64:. 11–25

EEA 2006. *Greenhouse gas emission trends and projections in Europe 2006*. Oct 2006. European Environment Agency: Copenhagen.

EAC (Environmental Audit Committee). 2007. *Pre-budget 2006 and the Stern Review*, Fourth Report of Session 2006–07, HC 227, Stationary Office, London.

Ecologist. 2007. The Stern Report: Editor's Comment. *The Ecologist*, December 2006/January 2007: 17–18.

Egenhofer, C., and L. van Schaik. 2005. *Towards a Global Climate Regime: Priority Areas for a Coherent EU Strategy*. Centre for European Policy Studies, Brussels.

Grubb, M. 2004. Kyoto and the Future of International Climate Change Responses: From Here to Where? *International Review for Environmental Strategies* 5 (1): 1–24.

Hajer, M. 1995. *The Politics of Environmental Discourse*. Oxford: Oxford University Press.

Helm, D. 2003. The Assessment: Climate Change Policy. *Oxford Review Of Economic Policy* 19 (3): 349–361.

Helm, D., C. Hepburn and R. Mash. 2003. Credible Carbon Policy, *Oxford Review Of Economic Policy* 19 (3): 438–450.

Hisschemoller, M., and R. Hoppe. 1995. Coping with Intractable Controversies: The Case for Problem Structuring in Policy Design and Analysis. *Knowledge and Policy: The International Journal of Knowledge Transfer and Utilization* 8, (4): 40–60.

131

IPCC. (Intergovernmental Panel on Climate Change). 2001a. *Climate Change 2001: Synthesis Report, Third Assessment Report of the Intergovernmental Panel on Climate Change*. Cambridge University Press, Cambridge.

IPCC. (Intergovernmental Panel on Climate Change). 2001b. *Impacts, Adaptation, and Vulnerability – Contribution of Working Group II to the IPCC Third Assessment Report*, edited by J. J. McCarthy, O. F. Canziani, N. A. Leary, D. J. Dokken and K. S. White. Cambridge: Cambridge University Press.

IPCC (Intergovernmental Panel on Climate Change). 2007a. *Contribution of Working Group I to the Fourth Assessment Report of the Intergovernmental Panel on Climate Change*. Summary for Policymakers, IPCC, Geneva.

IPCC (Intergovernmental Panel on Climate Change). 2007b. *Climate Change 2007: Impacts, Adaptation and Vulnerability Working Group II Contribution to the Intergovernmental Panel on Climate Change Fourth Assessment Report*. Summary for Policy Makers, IPCC, Geneva.

IPCC (Intergovernmental Panel on Climate Change). 2007c. Contribution of Working Group III to the Fourth Assessment Report of the Intergovernmental Panel on Climate Change, Summary for Policy Makers, IPCC, Geneva.

Jessop, B. 2003. Governance and Meta-Governance. In *Governance as Social and Political Communication,* edited by H. Bang. Manchester: Manchester University Press.

Jordan, A., and I. Lorenzoni. 2007. Is There Now a Political Climate for Policy Change? Policy and Politics after the Stern Review, *Political Quarterly* 78 (2): 310–319.

King, D. A. 2004. Climate Change Science: Adapt, Mitigate, or Ignore? *Science* 303 (5655): 176–177.

Kingdon, J. W. 1995. *Agendas, Alternatives and Public Policies* (2nd edn). New York: Harper Collins.

Kok, M., W. Vermeulen, A. Faaij and D. de Jager (Eds). 2002. *Global Warming and Social Innovation: The Challenge of a Climate-Neutral Society.* London: Earthscan.

Lenton, T. M., M. F. Loutre, M. S. Williamson, R. Warren, C. Goodess, C. M. Swann, D. R. Cameron, R. Hankin, R. Marsh and J. G. Shepherd. 2006. *Climate Change on the Millenial Timescale*. Tyndall Centre Technical Report 41.

Lomborg, B. 2001. *The Sceptical Environmentalist*. Cambridge: Cambridge University Press.

Olson, M. 1969. *The Logic of Collective Action: Public Goods and the Theory of Groups*. New York: Schocken Books.

Matthews, K., and M. Patterson. 2005. Boom or Bust? The Economic Engine Behind the Drive for Climate Change Policy. *Global Change, Peace and Security* 17 (1): 59–75.

McNeely, J. A. 2006. Biofuels: Green energy or grim reaper? BBC News, http://news.bbc.co.uk/1/hi/sci/tech/5369284.stm (last accessed 2 July 2007).

Meehl, G. A., W. M. Washington, W. D. Collins, J. M. Arblaster, A. Hu, L. E. Buja, W. G. Strand and H. Teng. 2005. How much more global warming and sea level rise? *Science* 307:1,769–1,772.

Meinshausen, M. 2005. What Does a 2°C Target Mean for Greenhouse Gas Concentrations? A Brief Analysis Based on Multi-Gas Emission Pathways and Several Climate Sensitivity Uncertainty Estimates. In *Avoiding Dangerous Climate Change*, edited by H. J. Schellnhuber. Cambridge: Cambridge University Press.

MNP 2007. China now no. 1 in CO_2 emissions; USA in second position, http://www.mnp.nl/en/dossiers/Climatechange/moreinfo/Chinanowno1inCO2emissions USAinsecondposition.html. Netherlands Environmental Assessment Agency (last accessed 20 June 2007).

Page, E. A. 2006. *Climate Change, Justice and Future Generations*. Cheltenham: Edward Elgar.

Pielke, R., G. Prins, S. Rayner and D. Sarewitz. 2007. Lifting the taboo on adaptation. *Nature* 445, 8 February 2007.

Pierre, J., and B. G. Peters. 2000. *Governance Politics and the State*. Basingstoke: Macmillan.

Rittel, H., and M. Webber. 1973. Dilemmas in a General Theory of Planning. *Policy Sciences*, 4, 155–169.

Sachs, W. 1999. *Planet Dialectics: Explorations in Environment and Development*. London: Zed Books.

Schellnhuber, H. J. (Ed.). 2005. *Avoiding Dangerous Climate Change*. Cambridge: Cambridge University Press.

Schon, D. A., and M. Rein. 1994. *Frame Reflection: Towards the Resolution of Intractable Policy Controversies*. New York: Basic Books.

Shue, H. 1993. Subsistence Emissions and Luxury Emissions. *Law & Policy*, 15 (1): 39–59.

Stern, N. 2007. *The Economics of Climate Change*. Cambridge: Cambridge University Press.

Sugiyama, T., K. Tangen, A. Michaelowa, J. Pan and H. Hasselknippe. 2003. *Scenarios for the Global Climate Regime*, Briefing Paper (http://www.fni.no/post2012/briefing-paper.pdf).

Toth, F., and M. Mwandosya. 2001. 'Decision-Making Frameworks'. In B. Metz et al. (Eds), *Climate Change 2001: Mitigation*. Contribution of Working Group III to the Third Assessment Report of the Intergovernmental Panel on Climate Change, Cambridge University Press, Cambridge.

Verweij, M., M. Douglas, R. Ellis, C. Engel, F. Hendriks, S. Lohmann, S. Ney, S. Rayner and M. Thompson. 2006. Clumsy Solutions for a Complex World: the Case of Climate Change. *Public Administration* Vol. 84, No. 4: 817–843.

Weale, A. 1992. *The New Politics of Pollution*. Manchester: Manchester University Press.

Wettestad, J. 2000. The Complicated Development of EU Climate Policy. In M. Grubb and J. Gupta (Eds), *Climate Change and European Leadership*. Dordrecht: Kluwer Academic Publishers.

World Public Opinion. 2007. *Poll Finds Worldwide Agreement That Climate Change is a Threat*, available at http://www.worldpublicopinion.org/pipa/articles/home_page/329.php

Young, O. 1989. *International Cooperation: Building Regimes for Natural Resources and the Environment*. Ithaca, NY: Cornell University Press.

Young, O. R. 2002. *The institutional dimensions of environmental change. Fit, interplay and scale*. Cambridge, MA: The MIT Press.

Notes

1. All facts contained in this chapter were correct at the time of writing.

2. See for example, the assessment of the prospects for the EU by the European Environment Agency (EEA 2006).

3. This international poll finds widespread agreement that climate change is a pressing problem.

4. The Kyoto Protocol recognizes six gases as contributing to the greenhouse effect: carbon dioxide (CO_2), methane (CH_4), nitrous oxide (N_2O), hydrofluorocarbons (HFCs), perfluorocarbons (PFCs) and sulphur hexafluoride (SF_6). Other significant contributors include water vapour.

5. China became the world's largest emitter of CO_2, much earlier than had been expected, in 2007 (MNP 2007).

6. Climate models project that the world is committed to a further warming of $0.5°–1°C$ over the next several decades due to past emissions (Meehl et al. 2005).

7. The Convention recognizes the climate system as a shared resource whose stability can be affected by industrial and other emissions of carbon dioxide and other greenhouse gases.

8. This consensus is manifest in the pronouncements of the Intergovernmental Panel on Climate Change (IPCC).

9. The Asia-Pacific Partnership for Clean Development and Climate, announced in July 2005, brought together Australia, China, India, Japan, South Korea and the USA, with the intention of developing technologies to reduce greenhouse gas emissions rather than setting specific targets to reduce them. The partnership was initiated by Australia and the USA, neither of which ratified the Kyoto Protocol.

10. It is important to recognize that even under a 'universal' approach, there can be scope for a range of parallel initiatives. Many sections in the UNFCCC text recognize the need for parallel activities to complement whatever actions might be undertaken within the framework.

11. The European Union is an interesting case where member states have agreed to 'pool' sovereignty in important policy areas. Even this relatively well-integrated governance system, however, proved unable to overcome objections to a common carbon energy tax (Wettestad 2000).

12. The Clean Development Mechanism was intended to help developing countries move on to more sustainable and lower-emitting paths of economic development without these countries themselves bearing the assumed costs.

13. Globally, governments spend around US $250,000m. a year on energy subsidies, of which over $80,000m. was in the OECD countries and over $160,000m. in developing countries (Stern 2006).

14. Limiting emissions such that the global temperature rises no more than 2–2.8°C may cost up to 3% of global GDP in 2030 (IPCC 2007c). This would mean emissions beginning to decline within around 15 years and being cut to less than 50% of 1990 levels by 2050. According to the IPCC, by 2050 the GDP cost of achieving this could be as much as 5.5%, though regional costs may differ significantly.

The Ethical Dimensions of Global Environmental Change

CHUKWUMERIJE OKEREKE

INTRODUCTION

Ethics does not have an exalted place in the global environmental change discourse. It is not one of the more common terms that one encounters in mainstream policy discussions on global environmental change. This is quite simply because global environmental problems are not, in many quarters, conceptualized in moral terms. That is, they are not perceived as having any direct link with, or necessitating serious discussions over, the moral standards that are prevalent in individual and public life. The more official approach is to characterize environmental problems in terms of a challenge to economics, law, and science and technology (see for example Stern 2007). For this reason ethics does not compete well with vocabularies like 'governance', 'management', 'regime', and 'cost-benefit' in most intergovernmental deliberations on the cause and cure of global environmental problems. Furthermore, as Shue (1995: 453) points out, ethics 'has had a bad reputation among many serious students of international affairs because they have thought that ethical considerations must be both individualistic and utopian'.

Notwithstanding, there is ample evidence that ethical discussions have played a considerable role in the development of environmental awareness both nationally and at the international level (Attfield 1999, 2003; Gleeson and Low 2001). It is also evident that ethical considerations have had significant influence on the development of the political positions of a wide range of entities, including states, environmental NGOs and corporations on the environment (Paterson 1996, 2001; Grubb 1995).

Ethics might not be a popular term in international affairs but it remains an inseparable aspect of every political process insofar as these demand choices among different ideas of what is right or desirable (Holden 1996: 4). Moreover, the 'global and systemic nature of many environmental problems' (Francis 2003:116) as well as the broader changes in the patterns of social economic relations among states have resulted in many groups with 'different cultures, values and loyalties' (Low and Gleeson 2001: 11) emerging as legitimate concerns and voices both in the interpretation and development of solutions for global environmental change (Dyer 1993). It seems imperative, therefore, that a meaningful approach to global environmental governance must be such that encompasses a broad range of these cultural and ideological perspectives.

Meanwhile, some ethical theorists insist that drastic changes in the global environment so unambiguously point to a fundamental limitation in the dominant philosophies that underwrite the relationship between human beings as well as their relationship to the environment and its non-human content (Eckersley 1992; Heyd 2003; Attfield 2003; Brennan 2003). Accordingly, they argue that there is little hope of finding a lasting solution for global environmental problems without ethical considerations being put at the centre of international affairs (Shue 1993, 1995; Jamieson 1992; Sachs 1999; Brown 2004). However serious 'issues' remain both in conceptual and pragmatic terms with respect to how the different and, often, conflicting ethical claims on the environment may be resolved.

This essay briefly examines the connections between ethics and the politics of global environmental change. Firstly, in the next section, I provide a brief discussion on the nature of environmental ethics and then highlight the link between ethics and environmental public policy. In section three, I contrast ethical conceptions of the environment with more conventional characterizations of the environmental challenge, in order to indicate some of the core issues and related questions about which ethical theorists engaging with the global environmental change discourse tend to be concerned. In section four, two out of the seven issues and related questions are discussed in more detail, as a means of highlighting the critical input and main arguments, as well as some of the limitations of environmental ethics discussions. Section five contains a brief discussion on the key challenge of ethics on global institutional governance of environmental change; followed, in the last section, by a brief concluding remark.

ETHICS, POLITICS AND PUBLIC POLICY

Ethical considerations can be described as centring on attempts to arrive at well-based practical moral standards that should guide individual or public life usually in terms of rights, obligations, benefits to society, fairness, or specific virtues such as honesty and prudence (Aristotle 1847; Mill 1893; Rawls 1971; Barry 1989). The primary aim is to provide us with guidance in choosing right from wrong and in imposing reasonable obligations and duties within a society mostly by seeking ways to resolve conflict over different conceptions of the good (MacIntyre 1981: 113). This usually involves, first, an acknowledgement of the presence of diverse and in some cases conflicting ideas of the good life. But then, critically, ethical discussions proceed to interrogate such various ideas and sentiments against the background of the more fundamental ethos and values upon which society ought to function in order to articulate what moral principle or system of principles should guide behaviour or policy.

Clearly, then, the bridge between ethics, politics and public policy is a very short one since the central concern of politics is the search for the most effective ways to authoritatively allocate values, power and resources in order to avoid conflict and maximize welfare within a political community (Leftwich 1984). Indeed neither politics nor public policy can be value-neutral; for, if people did not have different views on things, and different notions of what is valuable or preferable, there would be no tendency for conflict or the imperative to balance among different options or competing conceptions of the good, as is the case with much of politics and public policy.

ENVIRONMENTAL ETHICS

The focus of environmental ethics is to articulate what ought to be the moral relationship of human beings to non-human nature as well as the moral relationship among human beings in the context of their relations to the environment (Attfield 1991, 1999, 2003; Brennan 2003). The starting point is often taken to be the 'identification of those beings which are deemed to be morally considerable' (Hayward 1996: 55). This then leads to the discussion about what values and moral obligations are due to such beings.

It has to be accepted that many of the earlier forms of environmental ethics were clearly dominated by discussions about the moral value of non-human nature and the kinds of obligations and duties that are due them from human agents. These discussions were pursued without much reference to the socio-political dimensions of humans' relations to the environmental and mostly in the context of national rather than international politics. In other words, the earlier forms of environmental ethics discussions were largely silent on the 'connection between environmental destruction, unequal resource consumption, poverty and global economic order' (Brennan 2003: 22). As Alder and Wilkinson (1999: ix) correctly observe, these discussions were largely confined within a 'fringe area of philosophical exotica' without translating much to practical political questions (see also Heyd 2003: 9; Brown 2004: 111). Perhaps this almost exclusive preoccupation with the preservation of the wilderness, mountains and such other exotic places of recreation for the rich is one of the main reasons why environmental ethics failed to capture much attention from mainstream politics and environmental policy-making circles.

However, the 'imbalances in the basic approach' (Brennan 2003: 21) are being corrected with an increasing number of ethical discussions focusing on issues that are of fundamental practical importance in national politics and in the international relations of the environment (e.g. Shue 1992; Jamieson 1992; Paterson 1996; Grubb 1995; Attfield 1999; Page 2006, 2007; Dobson 1998, 2004). This newer approach has been especially facilitated by the incidence of complex environmental problems like climate change, which serves well in highlighting the ethical dilemmas implicated in the global environmental change and its governance (see Rayner and Okereke, in this volume). It is indeed probably fair to suggest that of all issues of international concern, it is the subject of global environmental change that has provided the impetus for the widest and most radical engagement with ethical discussions in international affairs over the last 10 years (see Okereke 2007). Even so, one would still have to admit that popular as it has become in the last two decades, the language of ethics has yet fully to penetrate mainstream environmental public policy circles.

As for the charge that ethical discussions must be utopian and individualistic, it would have to be conceded that ethical discussions, given their nature, would always tend to accommodate a considerable amount of abstract formulations that can make them seem idealistic. However, the ones that are worthy of our attention do not focus only on what ought to be; but equally consider what presently is. This is because effective guidance on how to attain an ideal usually requires mapping from the present position (Shue 1995: 454); but also because, as already indicated, ethical ideas form an important and integral constituent of existing contexts, structure and processes. Ideally then, there should be a 'duality of purpose; one which reflects the differences between 'is' and 'ought' while rejecting the old positivist notion of a logical gulf between them' (Jones 1995: 4).

ETHICAL VERSUS RATIONAL CONCEPTIONS OF GLOBAL ENVIRONMENTAL CHANGE

Conventional environmental policy studies, as noted, present environmental problems mainly as a challenge for science and technology, law and economics. In this view, 'global environmental change and its attendant consequences are seen as problems 'to be managed'' (Jamieson 1992: 143). The 'managerialist' or 'technocratic' approach to global environmental change (Pepper 1984) has as its starting point the conceptualization of environmental problems in terms of economic externalities or as collective action problems inherent in the management of public goods.

As a problem of economic externality, environmental degradation is seen as arising because the social cost of production is not usually reflected in the market cost of goods (Anderson and Leal 1991; Common 2003). The argument is that the differential between the social cost and private cost of goods provides a negative incentive for polluters to change to more environmentally benign behaviours. Similarly, as a collective action problem, drastic environmental degradation is viewed as resulting from the fact that rational self-interested entities (individuals, states, companies, etc.) tend to choose environmentally detrimental behaviours because of short-term gains, and regardless of the knowledge that such choices would result in the worst outcome for everyone in the long run.

Either way, the solutions to environmental problems are said to lie in the elaboration of effective regimes to regulate behaviour and allocate property rights as well as in the internalization of environmental costs by creating a more perfect market (Anderson and Leal 1991; Beckerman 1994; Common 1996; Stern 2007). Meanwhile, innovations in science and technology are desperately sought so that the society could afford to produce and consume more resources while generating less waste (Weale 1992; Mol 2001). In the international arena, achieving these goals is thus seen as the primary objective of international environmental co-operation (cf. Young 1989).

Critical ethical analysis, however, takes a different approach. It considers that drastic changes in the global environment raise fundamental questions about the dominant modes of relationship between human beings as well as their relationship to the environment and its non-human content (Goodin 1992; Hayward 1998; Low and Gleeson 2001; Dobson 1998, 2003, 2004). Arising form this conceptualization, it is posited that a more viable solution to environmental problems (nationally and globally) must entail some re-examination of and changes in the values that underpin intrahuman relations and humans' relations to the natural environment. This perception is well captured in the words of the authors of the highly influential book *The Limits to Growth*: 'We affirm finally that any deliberate attempt to reach a rational and enduring state of equilibrium by planned measures rather than by chance or catastrophe, must ultimately be founded on a basic change of values and goals at individual, national and world levels' (Meadows et al. 1972: 45).

To be clear, environmental ethics perspectives do not generally deny the importance of economic information or effective legal frameworks in environmental decision-making. Nor is it suggested that rational self-interest is not a strong motivation for individual and public action. However, ethical perspectives challenge the prevalent 'more grandiose claim' (Jamieson 1992: 143) that economics provides about the most important information for environmental policy decision-making. Indeed it is contended that an exclusively managerialist approach to the environmental question is bound to fail in that such an approach avoids a critical engagement with the underlying values and systems—e.g. greed, consumerism, unequal resource distribution, unfettered capitalism and economic liberal individualism, etc—that are at the heart of current drastic changes in the environment (cf. Daly 1992; Jamieson 1992; Gardiner 2004; Vanderheiden 2005).

At the same time, contrary to the liberal institutionalist assumption that international co-operative institutions work to protect the mutual interest of all negotiating parties, ethical perspectives point out that legal and co-operative arrangements can and often do mask important disagreements over what policies ought to be pursued and who should participate in the decision-making procedures (Chatterjee and Finger 1994; Low and Gleeson 2001; Okereke 2006). Furthermore, it is pointed out that the rule-content of co-operative arrangements is by no means value-neutral but rather that co-operative rules and regulations often reflect preferences for some values over and above contending alternatives (Hurrell 2001; Bernstein 2001; Okereke 2007). For example, Bernstein's (2001) analysis of global institutions for environmental co-operation reveals that despite a serious attempt to portray these institutions as working to secure global environmental sustainability, they actually serve for the most time to promote a specific socio-economic order that is normatively contrary to the ideals imbedded in the notion of environmental sustainability. Similarly Okereke (2006, 2007) has shown that while there seems to be a common agreement that international environmental institutions ought to be fair and equitable to be accepted, the actual notions of justice that underpin global environmental regimes can hardly be said to be either compatible with the ideal of sustainability or responsive to the concern of the majority of the world's population.

The argument for the interrogation of the values that underpin public life or co-operative institutions does not equal a proposition that all options should be given equal weighting in decision-making. But it does imply, first, that there is no prima-facie reason why all options should not be considered; and, secondly, that the process of justifying, selecting or rejecting one value above possible alternatives must be clear and fair. The reason is quite simply that in the absence of such discussions, there is a serious possibility that the rules, policies and modalities of institutions of co-operative arrangements for response to global (environmental) challenges may very well be skewed in the favour of those with greater power and resources (see Shue 1992; Sachs 1999; McCarthy and Prudham 2004). Hence, a major consideration of ethical theories is the fairness of co-operative arrangements and the distributional outcome of rule-based institutions.

The following, in general, constitute the main (overlapping) issues and related questions about which environmental ethical theories concern themselves (Brown et al. 2006). Their main contribution to the global environmental change discourse equally lies in their attempt to highlight the importance of these issues and to engage with the complex challenges they pose to the conventional approaches to environmental public policy and international co-operation.

1. Allocation of value: Do only humans have ultimate value or does the non-human nature qualify for moral considerations? How may we justify our obligations towards the non-human? How do we deal with the many conflicts that arise when we grant any non-humans moral standing?
2. Distribution of resources: Which humans matter in the allocation of environmental resources? Only those in our nation? Only those in our region? All in the globe? Only those alive now or including members of future generations? What procedure for distribution should we follow? Who has the right to participate in the decision-making process?
3. Nature of obligation: What obligations do we owe other humans? Is there a right to a healthy environment? One that provides for a full range of needs? Of opportunities too? How do we define this in specific policy areas?
4. Responsibility for past damages: Who is responsible for the consequences of past damages to the environment?
5. Dealing with uncertainty: How do we deal with uncertainty and complex natural systems? What is the ethical significance of making critical decisions in a condition of scientific uncertainty?
6. Environmental obligations: How do we think of environmental obligations? In cost-benefit terms? Based on other values? How do we balance among competing interests?

In the next two sections, I will discuss some aspects of the first two of these issues/questions in some more detail. Space constraints prohibit the discussion of all six issues and even with respect to the two questions discussed there are a number of angles that will not be fully explored. My attempt will be to highlight the evolution of thought as well as some of the key themes running through the recent debates. The two issues are selected because of their significance in terms of the development of the environmental ethics research agenda, and also because they actually provide space to capture some of the concerns implicated in the other questions.

THE CASE FOR ENVIRONMENTAL MORALITY

A key dimension of environmental ethics, as already noted, involves discussions on the moral status of the non-human environment and by implication the moral obligations of human beings to members of other species on earth. Questions about the moral worth of the non-human environment are not new; but rather form an important element of traditional Western philosophical thought. For example, the writings of Aristotle (*Politics*, Book 1, and Chapter 8), Thomas Aquinas (*Summa Contra Gentiles*, Book 3, Part 2) and Immanuel Kant (*Lectures on Ethics*) all contain explicit contemplations on humans' moral relationship with nature albeit in support of anthropocentrism (see Attfield 1991, 1999). Similarly, most of the world's religions embody considerable stipulations and principles on how man should relate to nature (see Palmer and Finlay 2003). However, it was only with the emergence of the new wave of environmentalism that more explicit attempts to articulate clear and coherent positions on this became more pronounced and widespread.

The starting question is usually posed in terms of whether non-human things have *intrinsic* or merely *instrumental* value. To say that they have only *instrumental* value implies a conviction that they have, in themselves, no independent moral worth. In other words, that their value is essentially determined by their usefulness to human beings. This then leads to the position that they are not subject to moral considerations from humans (anthropocentrism). If, on the other hand, they are considered as having *intrinsic* value, it means that they are deemed as having a moral worth of their own accord which is not dependent on the (instrumental) value attached to them by human beings. This position implies that there is 'a *prima facie* direct moral duty on the part of moral agents to protect it or at least refrain from damaging it' (Brennan and Yeuk-Sze 2002) (non-anthropocentrism).

Some environmentalists argue that humans alone should be the right focus of morality while others consider that 'it is arbitrary, if not obnoxious to confine moral status to human beings' (Alder and Wilkinson 1999: xi). Because the onus of proof is usually on non-anthropocentric writers to show why non-human nature should be accorded moral consideration, I will limit myself here to discussing their key arguments as well as some of the main limitations.

Those who consider that non-human nature should be a subject for moral consideration do so mainly by developing various arguments against the more dominant anthropocentric philosophy. First and foremost they seek to draw a link between anthropocentrism and drastic changes in the natural and human environment (White 1967; Bookchin 1980; Naess 1989; Coward 2006). Here it is argued that anthropocentrism is predicated on human-centred self-indulgence, which in turn leads to irresponsible exploitation of natural resources. In this perspective, what is needed to preserve the natural environment is a rejection of the irreverent, self-indulgent and consumerist mindset that is deeply ingrained in the anthropocentric worldview. Secondly, a variation, often described as 'social ecology', seeks to establish a connection between man's domination of nature, on the one hand, and political control, social injustice and human oppression on the other hand. Routley and Routley (1980), Kelly (1990) and Bookchin (1990, 2005), for example, have all argued that anthropocentrism is intimately connected to class chauvinism and racism, all of which entail unjustifiable discrimination against those outside the privileged class. Moreover it is argued, here, that in regarding nature as a mere resource for man, humanity is missing a wonderful opportunity for developing its more aesthetic and moral 'inner nature' faculties which are best discovered via a harmonious 'Romantic' relationship with nature. At the same time, some ecofeminists have suggested that there is a close connection between anthropocentrism and the oppression of women since anthropocentrism is chiefly characterized by the creation of dualism and the logic of control deeply implicated in patriarchal modes of thinking (Mies and Shiva 1993; Plumwood 1993, 2002). Altogether, it is argued that a lasting solution to the problem of drastic environmental change lies in the promotion of a non-anthropocentric, more holistic ethic which emphasizes the unity of all beings and acknowledges the moral worth of the non-human environment.

These positions as well as their many other variations have altogether had a significant impact on the development of environmental awareness both nationally and internationally. Critically they played important roles in the formation of many of the green parties across Europe in the 1970s and 1980s (Doherty 2002; Doherty and Doyle 2006). Currently, several countries in Europe have vibrant green parties, some of which command considerable influence in the national parliament (Müller-Rommel 2002). Also, these green parties are fairly well represented and exert considerable influence in the European Parliament.

In the USA and Canada, there is evidence that the Wilderness campaign organizations have in the past exerted and continue to exert notable influence in the development of public land laws and general environmental policy processes (Niemeyer 2004). Moreover some of the incredibly influential multinational NGOs, such as the World Wildlife Fund (WWF) and the World Conservation Union (IUCN), have strong historical links with these ideas. It should be noted that even the UN in 1982 adopted a World Charter for Nature Resolution which reads in some parts as a statement on non-anthropocentrism. The UN document for example expresses the conviction that 'every form of life is unique warranting respect regardless of its worth to man'. The document equally calls for the elaboration of a 'moral code' to guide man in his relationship with non-human nature. It is indeed fair to suggest that the very concept of sustainable development actually developed through these non-anthropocentric discourses.

However, despite their notable influence in developing the environmental agenda as well as commanding significant emotive appeal, descriptive arguments against the anthropocentric philosophy do not in themselves alone suffice as ethical justifications for non-anthropocentrism. In order to show that non-human nature qualifies for inclusion as a subject for moral consideration, more principled and theoretically coherent arguments are needed. However, it is precisely at this juncture that important differences emerge. There are, to oversimplify, at least four main approaches, each of which has a number of serious drawbacks.

Consequentialist defence of non-anthropocentrism draws mainly from utilitarianism and starts from the assumption that the ultimate criterion for morality is whether an action produces pain or pleasure. Actions that produce pain are bad while those that produce pleasure are good and worthy of pursuit. This leads to the suggestion that only the beings that are capable of feeling pain or pleasure are the ones that qualify for moral consideration (Singer 1975, 1993). The problem with this view, however, is that it excludes important areas of concern for environmentalists, including landscapes, plant species, rivers and biotic communities. Moreover, this approach is not internally coherent. Because it values human pleasure over and above animal pleasure, it winds up sanctioning aspects of animal cruelty (e.g fox hunting) insofar as these bring pleasure to a large percentage of the human population. It is thus 'unclear to what extent a utilitarian ethic can also be regarded as an environmental ethic' (Brennan and Yeuk-Sze 2002).

A deontological ethical approach opens space for the inclusion of other aspects of non-human nature, including trees, based on the argument that neither their existence nor their internal qualities are owing to man's creative efforts. The straightforward argument that it is not morally right to eliminate what you cannot create is forceful. But if all organisms are in themselves inclusive 'centres of life' and thus worthy of moral consideration, it means that there is no prima-facie justification for regarding a human being as more valuable than grasses, trees, bacteria or molluscs. But this reasoning would seem to be very counterintuitive (cf. Attfield 1995; Agar 1997; Curry 2005; Evans 2005).

A significant strand, which also draws upon deontological theories, contends that it is the whole biotic communities rather than the individual species that are subjects of moral consideration (Leopold 1949; Rolston 1989). But then the *land ethic* as it is often called leads to a direct conflict with animal rights groups in that it seems to support the idea that individual animals could be sacrificed for the well-being of the biotic community.

The last is the virtue-based approach, which draws mainly from Aristotle's *Nicomachean Ethics* (see O'Neill 1992, 1993; Barry 1999). In contrast to the consequentialist and deontological approaches, which stress 'pleasure' and 'rights', virtue ethics focuses on individual attitude and the underlying motivations for action. The central thesis is that 'the outcomes of our actions are less important than that we have the right attitude and do things for the right reasons' (Alder and Wilkinson 1999: 42). Virtue ethics promotes attitudes such as care, humility and moderation, which are argued to be both foundational in living a morally flourishing life but also as promoting a more wholesome relationship with the non-human world (Barry 1999; Alder and Wilkinson 1999). Also, in emphasizing prudence and caution, virtue ethics may provide a better base for important environmental concepts like the precautionary principle as well as for dealing with issues of scientific uncertainty that are of great significance in the context of climate change (see Jamieson 1992; Barry 1999). However, virtue ethics' defence of (human) nature is only derivative since its central focus is on encouraging the flourishing of human life. To be internally coherent, virtue-based arguments must ultimately be anthropocentric.

The consequences of a human-centred life and reckless attitude to nature are self-evident but it is an uphill task to articulate a coherent theory that provides a warrant for the moral considerability of non-human nature. Given this dilemma, some have suggested that what is needed after all is not a total rejection of anthropocentrism but an embrace of some form of cautious anthropocentrism or of what Brennan calls 'enlightened anthropocentrism' (Brennan 2002). The argument is that since it seems that the moral duties we have towards the environment ultimately derive from human-centred interests, all that is needed is prudence in our dealing with nature. A prudent approach to nature is thought to be 'sufficient for practical purposes and perhaps even more effective in delivering pragmatic outcomes, in terms of policy-making, than non-anthropocentric theories' (Brennan and Yeuk-Sze 2002). Moreover, as eloquently argued by Guha (1989), there is a real danger that the promotion of certain versions of non-anthropocentrism, such as Romanticism, might only serve the purpose of securing exotic recreational places for the rich élites while taking the focus off the issues of poverty, unequal resource control and distributional justice which are the main environmental concerns of the majority of the world's population.

JUSTICE AND THE ENVIRONMENT

The other main dimension of environmental ethics I discuss in the essay relates to questions about the nature and extent of linkage between distributional justice and the idea of environmental sustainability. This involves concerns over what should be the main principle for the allocation of environmental resources both nationally and internationally and how to resolve the complex issues relating to the justice of allocation and distribution between the present and future generations of human beings (Dobson 1998; 1999; Page 2006, 2007).

Mainstream environmentalism did not develop with adequate concern for distributional justice as much of the focus was on conservation and the protection of endangered non-human species (Pepper 1986; Bowler 1992; Okereke 2007). Indeed, for a very long time, the environmental question was regarded as a special concern for ecologists and evolutionary biologists rather than an issue for politics and social struggle (Bowler 1992; Pepper 1993). It was actually at the instance of attempts to develop international co-operative arrangement for the management of the global environment that questions of distributional justice became firmly introduced into the environmental discourse. The term 'environmental justice' itself was, however, eventually popularized and has since become somewhat closely associated with the environmental justice struggles in the USA (Kutting 2004: 115).

At the international level, the link between justice and the environment has been quite unsurprisingly pushed mostly by those from the developing countries and a number of their supporters from the West (Agarwal and Narain 1991; Shue 1992; 1993; Hurrell 2001; Parks and Roberts 2006). There are mainly four key moments in history that provided the impetus for the development of the justice environment argument. The first was the long-drawn attempt to elaborate the third UN Law of the Sea (UNCLOS III) (1967–82). During this process, the developing countries made the point that the old customary law of the sea was essentially designed to secure the slave-trading and colonial interests of the great world powers while at the same time serving to provide the same group with a cheap dumping ground for their industrial wastes (Sanger 1986; Freidheim 1993). They subsequently called for an end to an ocean regime that provides for the wanton and inequitable exploitation of the common resources of the sea by those having the technological and economic advantage (Pardo 1975; Ramakrishna 1990a; Vogler 2000). This view was eventually reflected in the document of the new regime which states that the new law of the sea is intended to 'contribute to the realization of a just and equitable international economic order [and] which takes into account the interests and needs of mankind as a whole and, in particular, the special interests and needs of developing countries' (UNCLOS III Preamble).

The second critical moment was the first UN Conference on the Human Environment (UNCHE) in 1972. In the preparations leading to the conference, the developing countries were very vocal in stating their conviction that questions of distributional justice intertwine and therefore cannot be separated from the global environmental debate (Founex Report 1971). Indeed they made it very clear that in their view, 'the major problem that would arise' in the search for global co-operation for environmental change was how the issues of distributional justice implicated in the whole process might be resolved (Founex Report 1971: 3). It was in response to these sentiments that the world leaders in UNCHE adopted the concepts of '*additionality*' and '*compensation*' as *key* principles in North-South global environmental co-operative efforts. These require, first, that developing countries should be provided with additional funds for the promotion of environmentally benign development and, secondly, that less industrialized countries should be rewarded for major dislocations of exports arising from the new emphasis on the environment (Ramakrishna 1990b: 429).

The other two moments include the development of the report of the World Commission on Environment and Development (the Brundtland report), published in 1987, and the UN Conference on Environment and Development (UNCED) held in Rio de Janeiro, Brazil, in 1992. Indeed by 1987 the link between international distributional justice and the environment had become so clear as to warrant the categorical statement by the Brundtland Commission that 'inequality is the planet's main environmental problem' (WCED 1987: 6) and that it would be 'futile to attempt to deal with the environmental problems without a broader perspective that encompasses the factors underlying world poverty and international inequality' (WCED 1987: 3). These convictions were later echoed in several documents agreed in the UNCED in 1992, including a statement that 'eradicating poverty and reducing disparities in worldwide standards of living are *indispensable* for sustainable development'.

In general, the demands for justice in the international environmental arena have taken the form of one or a combination of the following. The first is contestation for equal access to common property resources, which is linked with the argument that certain aspects of nature, such as the oceans, the deep seabed and Antarctica, belongs to all members of the world in common (see Sanger 1986; Vogler 2000; Thompson 2000). The second is the demand for equitable and fairer representation in global environmental decision-making circles, which stems from the observation that not all those affected by environmental problems have equal say in the elaboration of solutions to address such problems (see Achterberg 2001; Brown et al. 2004). The third is a clamour for fairer international economic structures and policies given the understanding that the present order works to perpetuate poverty and overdependence on scarce natural resources in some parts of the world while encouraging a profligate lifestyle in the other parts (Daly 1994; Martínez-Alier 2002; Clapp and Dauvergne 2005; see Paterson in this volume). Other forms of environmental justice include claims of rights to certain basic environmental goods such as land and clean water; and the demand for compensation arising from historical injustices, especially in relation to the past use of common (commonly-owned) resources (see Sachs 1999).

On the other hand, the developed countries sometimes argue that one of the greatest causes of global environmental degradation is rapid growth in human population. They suggest that the developing countries should as a matter of justice and environmental sustainability control their populations (cf. Hardin 1974; Lucas and Ogletree 1976). Some others argue that the main reason for poverty in the developing countries is lack of fiscal discipline and suggest that in demanding the transfer of resources from the North, the developing countries are basically asking for a right to steal (cf. Singer 1992).

As already mentioned, there have been equally notable attempts to establish a justice-environmental connection at continental, national and local levels. The most prominent example of these is of course the environmental justice movements in the USA where 'people of color' have sought to establish a link between environmental degradation and wider societal injustice, class domination, and political economic subjugation by the white-dominated élite (see Schlosberg 2005; Bullard 1994). The other critical dimension—which seems to occupy a priority position in the justice environmental discourses among Western scholars—is the question of whether future generations qualify as a reasonable community of justice and what should be the 'specification of a currency of advantage' that can be used to evaluate distributive outcomes across time' (Page 2007: 453; cf. de Shalit 1994; Dobson 1998; Page 2006).

Now, the question that is more or less implicit in these environmental justice struggles is about what principle of justice should guide societies in the allocation of environmental goods (and bads). The starting point is usually Dobson's (1998) observation that there are various conceptions of justice as well as different notions of sustainability, all of which correlate differently. Dobson argued that it is possible to imagine an environmentally sustainable society that is characterized by pervasive social injustice, just as it is possible to conceive a politically just society purchased 'at the cost of a deteriorating environment' (Dobson 1998: 3). He therefore called for works that seek to establish the notions of justice that might be considered compatible with the ideal of environmental sustainability both nationally and at the international level. Subsequently, the search has typically followed the path of appraising traditional theories of justice to explore the areas of convergence and divergence.

The utilitarian idea, which equates justice with maximizing happiness for the greatest number of people (Singer 1992), might be deemed consistent with the ideal of eliminating poverty world-wide. But utilitarianism would appear to directly sanction unsustainable exploitation of resources insofar as this would lead to happiness maximization. Besides, there are huge disagreements over what counts as utility and over how conflicting notions of utilities may be reconciled without sacrificing the genuine interests of the less privileged in society (cf. Paterson 1996). Justice as liberal egalitarianism, as developed by philosopher John Rawls, is seen as a corrective to utilitarianism since it is sensitive to individual preferences and equally seeks to secure the welfare of the least advantaged in society (Rawls 1971). However, there are concerns that the wide plurality in the conception of the good implied in liberal egalitarianism conflicts with the idea of environmental sustainability which would seem to sanction placing a definitive value on the environment (Sagoff 1995; Okereke 2007). What is more, Rawls himself rejects that his idea of justice should be applied in the international arena.

Libertarian ideas of justice regard the protection of property rights as the ultimate ideal (Nozick 1974; Hayek 1960). In this view protection of property rights is linked with protection of individual liberty and freedom. Further, the right to property and free trade are regarded as inviolable. But while this idea of justice might seem to encourage enterprise and maximization of individual liberty, it equally seems destined to lead to the endless sacrifice of the environment on the altar of individual or societal material greed. Besides, it has been shown that the pursuit of property right protection as sanctioned by libertarianism would ultimately deny many in the societies and especially future generations the opportunity to lead decent lives (see Dobson 1998; Kymlicka 2002).

Some have argued that the idea of justice most consistent with the idea of sustainability is the idea of justice as meeting needs (Langhelle 1999, 2000; Dobson 1998; O'Neill 1997). Most of these scholars draw inspiration from the Brundtland report, which defines sustainable development as development that meets the needs of the present generation without compromising the ability of future generation to meet their needs. However, the articulation of what constitutes a need remains a complex ethical theoretical challenge.

Besides the mainstream theories of justice, other principles of allocation have been contemplated in the international arena, especially with respect to the issue of climate change (see Rayner and Okereke in this volume). Debates on how to distribute emissions quotas among the nations of the world; how to deal with the issue of responsibility for historic greenhouse gas emissions; as well as how to deal with the huge differentials in vulnerability to the consequences of global climate change have thrown up consideration of several ideas of justice even in mainstream discussions on climate change policy. Some of the principles often considered include per caput, land mass, equality, willingness to pay, and grandfathering (cf. Grubb et al. 1992; IPCC 1996, 2001, 2007).

These debates are extremely sensitive because whereas their starting point might be the distribution of environmental benefits and burdens, they often fuse with deeper philosophical issues, such as the value of life, the desirability of forced population control, the responsibility for historical damages to the environment and the justice of present international political structures and institutions.

For example, discussions about the principle of allocation of carbon dioxide emissions under the climate change regime have led to some proposals that value life according to national GDP, which then means that one European is deemed as equivalent to 10 Chinese (cf. Nordhaus 1991; Fankhauser and Tol 1997). Similarly, the belief that the issue of international environmental justice can only be resolved if there is an authoritative supranational authority that can oversee the effective redistribution of global resources is one of the reasons why some have called for the establishment of a world government or at least a world organization for the environment. But there remain serious issues over how such a body, especially a world government, might be worked out in practice.

ENVIRONMENTAL ETHICS AND GLOBAL ENVIRONMENTAL GOVERNANCE

Before concluding, let me very briefly highlight one burning question that arises out of the foregoing discussion. The question is: to what extent is the current global environmental governance structure sufficiently imbued with the elements needed to entertain and respond to the challenges posed by environmental ethics? This question is pertinent because, as pointed out from the outset, normative discussions are not particularly popular in international affairs, which tend to assume economistic utilitarianism. The dominant characterization is that the international arena is an essentially anarchic environment without any central authority to impose on states. Moreover, rational self-interest is held as the key motivation of states in developing international co-operative arrangements. International affairs are thus deemed as responsive only to power and material calculus rather than to 'trifling' moral concerns. Furthermore, the state-centric logic of the international structure implies that many of the diverse views within states hardly get attention at the international level.

If, however, it is accepted that the moral questions implicated in the environmental discourse are valid as the discussions in the preceding sections would suggest, it should follow that the ethical basis of global environmental governance arrangements must be urgently 'held up for scrutiny' (Achterberg 2001: 183). Unfortunately, however, the ethical dimensions of global environmental governance arrangements remain one of the least discussed, both in literature and environmental policy circles. As Low and Gleeson (2001: 2) observe, 'the discourse on environmental ethics has tended to avoid the issue of global governance, whilst ethical issues have only been weakly developed in debates on international environmental regulation'.

While there are many important aspects, one of the critical issues is the apparent need to democratize current governance processes (implicit in Agenda 21) in order to give individual communities and groups that are directly affected by environmental problems the chance to have a say in discussing these problems and proffering possible solutions. It is clear that one of the most distinctive features of the environmental challenge is that nearly every individual human and almost certainly every community has either the ability to affect or stands to be affected by the action or inaction of others. This means that every human individual is a relevant stakeholder and needs to be provided with a fair opportunity to participate in the decision-making procedure. This proposition, with the exception of a few ecoanarchist voices, is not usually meant as a vote of no confidence in the current state system *per se*. But it does suggest that there is a need for a radical reform of the existing governance structures in ways that might allow real democratic politics to flourish at the international level and that nation states need not be the only legitimate players under the new arrangement.

Indeed, one of the curiosities of the modern political order is that the international system has somehow managed to remain somewhat insulated from the wave of democratization that has swept national political institutions even in the context of pervasive interdependence and a much-vaunted weakening of states' power due to socio-economic globalization. It is also curious that while there has been an immense proliferation in the number and types of agency beyond the state, especially with respect to global environmental governance—which at least in part indicates there is considerable political space between the national and the global—the idea of conferring legitimacy on entities other than states has not been squarely addressed. This might be because these entities—NGOs, indigenous community groups, world religious bodies, etc.—which mostly provide outlets for alternative voices, including 'voices from the margins' (Elliot 1997)—might themselves be in dire need of democratization.

Some have suggested that what is needed is not so much the incorporation of agencies beyond the state in governance as to ensure that the international system as presently constituted provides a level playing field for all states. But then again, in a system where some countries are of the size of a small municipality in other countries, it is a valid ethical question whether the views of all countries should be accorded equal weight. There is indeed no simple solution but what is required is that these ethical issues be taken up and given adequate attention. Perhaps if we start taking seriously the idea that every single individual (and not just states) has a stake in the environment, progress in the search for ways in which such interests might be effectively represented in decision-making circles would follow.

CONCLUSION

The essay has attempted to highlight some of the main issues that are more or less explicitly addressed in the ethical dimensions of the global environmental change discourse. The link between ethics, politics and public policy as well as the major ethical questions that arise in the discussions of the cause and cure of global environmental problems have been underscored. Using two of these issues/ questions—the moral status of the non-human nature and the right principle for the distribution of environmental benefits and burdens—the essay has attempted to identify some of the key contributions of ethical discussions to the development of environmental awareness and to the search for credible solutions. Despite the brevity of the above discussions, the essay makes clear that ethical and moral issues/questions cannot be regarded as exotic add-on extras designed to spice up more serious economic and legal arguments over how the environment might be saved. But, rather, that these issues have played considerable roles in generating the present-day level of global environmental consciousness and would seem to remain central in the search for possible solutions for coping with global environmental change. Despite, however, the seriousness of the ethical questions implicated in global environmental challenges, it remains the case that the language of ethics has not sufficiently penetrated mainstream policy-making circles and the current institutional governance structure does not seem adequately imbued with the elements needed to respond to the challenges they pose. Economics and science and technology will continue to be important tools in dealing with environmental problems both nationally and globally. However, without questioning some of the more fundamental values that underpin individual and societal life; and without the democratization of current global institutional arrangements, there is little chance that fair, equitable and lasting solutions of many of the current environmental challenges will ever be found. On the other hand, environmental ethics scholars still have the challenge to articulate more clearly the practical policy implications of some of the philosophical arguments that often characterize ethical discussions and how the conflicts generated by diverse ethical views may be resolved.

Bibliography

Achterberg, W. 2001. Environmental Justice and Global Democracy. In *Governing for the Environment: Global Problems, Ethics and Democracy*, edited by B. Gleeson and N. Low, pp. 183–195. Basingstoke: Palgrave.

Agar, N. 1997. Biocentrism and the Concept of Life. *Ethics*, 108: 147–168.

Agarwal, A., and S. Narain. 1991. *Global warming in an unequal world.* Centre for Science and Environment, New Delhi.

Alder, J., and D. Wilkinson. 1999. *Environmental Law And Ethics*. Basingstoke: Macmillan.

Anderson, T. L., and D. R. Leal. 1991. *Free Market Environmentalism*. Oxford: Westview Press.

Aquinas, T. 1975. *Summa Contra Gentiles*, tranlated by V. J. Bourke. London: University of Notre Dame Press.

Aristotle 1847/1998. *Nicomachean Ethics Book IV*, translated by J. A. K. Thompson.

Attfield, R. 1991. *The Ethics of Environmental Concern* (2nd edn). Athens: University of Georgia Press.

1995. *Value, Obligation, and Meta-Ethics*. Amsterdam/Atlanta: Editions Rodopi B.V.

1999. *The Ethics of the Global Environment*. Edinburgh: Edinburgh University Press.

2003. *Environmental Ethics: An Overview for the 21st Century*. Cambridge: Polity Press.

Barry, B. 1989. *Theories of Justice*. London: Havester-Wheatsheaf.

Barry, J. 1999. *Rethinking Green Politics*. London: Sage Publications.

Beckerman, W. 1994. Sustainable Development: Is it a Useful Concept? *Environmental Values* 3, (3): 191–210.

Bernstein, S. 2000. Ideas, Social Structure and the Compromise of Liberal Environmentalism, *ECPR*, 6 (4): 464–512.

Bernstein, S. 2001. *The Compromise of Liberal Environmentalism*. New York: Columbia University Press.

Bookchin, M. 1990. *The Philosophy of Social Ecology*. Montreal: Black Rose Books.

1980. *Towards an Ecological Society*. Montreal: Black Rose Books.

2005. *The Ecology Freedom: the emergence and dissolution of Hierarchy*. Edinburgh: AK Press.

Bowler, P. 1992. *The Fontana History of the Environmental Sciences*. London: The Fontana Press.

Brennan, A., and L. Yeuk-Sze. 2002. Environmental Ethics. *The Stanford Encyclopedia of Philosophy*, Edward N. Zalta (Ed.), http://plato.stanford.edu/archives/sum2002/entries/ethics-environmental.

Brennan, A. 2003. Philosophy. In *Environmental Thought*, edited by P. A. Edward and J. Proops, 15–33. Cheltenham: Edward Elgar.

Brown, D. A. 2004. Environmental Ethics and Public Policy. *Environmental Ethics*, 26: 110–112.

Brown, D. et al. 2006. *White Paper on the Ethical Dimensions of Climate Change*. Rock Ethics Institute Secretariat: Penn State University.

Bullard, R. D. 1994. *Dumping in Dixie: Race, Class, and Environmental Quality* (2nd Edn). Westview Press, Boulder, Colorado.

Chatterjee, P., and M. Finger. 1994. *The Earth Brokers: Power, Politics and World Development*. London: Routledge.

Clapp, J., and P. Dauvergne. 2005. Paths to a Green World: The Political Economy of the Global Environment. Cambridge, MA: MIT Press.

Common, M. 2003. Economics. In *Environmental Thought*, edited by P. A. Edward and J. Proops, 78–101. Cheltenham: Edward Elgar.

1996. *Environmental and Resource Economics: An Introduction*. Essex: Longman.

Coward, M. 2006. Against Anthropocentrism: the destruction of the built environment as a distinct form of political violence. *Review of International Studies* 32 (3): 419–438.

Curry, P. 2005. *Ecological Ethics: An Introduction*. Cambridge: Polity Press.

Daly, H. 1994. Fostering environmentally sustainable development: four parting suggestions for the World Bank. *Ecological Economics* 10, 183–187.

1992. Free market environmentalism: turning a good servant into a bad master. *Critical Review*, 6: 171–184.

de Shalit, A. 1994. *Why Does Posterity Matter?* London: Routledge.

Dobson, A. 1998. *Justice and the Environment: Conceptions of Environmental Sustainability and Dimensions of Social Justice*. Oxford: Oxford University Press.

2003. *Environmental Citizenship*: Oxford: Oxford University Press.

Doherty, B. 2002. *Ideas and Actions in the Green Movement*. London: Routledge.

Doherty, B., and T. Doyle. 2006. Beyond Borders: Transnational Politics, Social Movements and Modern Environmentalisms, *Environmental Politics*, 15 (5): 697–712.

Dyer, H. C. 1993. EcoCultures: Global Culture in the Age of Ecology. *Millennium: Journal of International Studies*, 22 (3): 483–504.

Eckersley, R. 1992. *Environmentalism andPolitical Theory: Towards an Eco-centric Approach*. London: UCL Press.

Elliott, L. 1997. *The Global Politics of the Environment*. London: Macmillan Press.

Evans, J. C. 2005. *With Respect for Nature: Living as Part of the Natural World*. Albany, NY: State University of New York Press.

Fankhauser, S., and R. S. J. Tol. 1997. The Social Cost of Climate Change: the IPCC second assessment report and beyond, *Mitigation and Adaptation Strategies for Global Change*, 1: 385–403.

Francis, L. P. 2003. Global Systemic Problems and Interconnected Duties. *Environmental Ethics* 5: 115–128.

Friedheim, R. L. 1993. *Negotiating the New Ocean Regime*. USA: South Carolina Press.

Gardiner, S. M. 2004. Ethics and Global Climate Change. *Ethics* 114: 555–600.

Gleeson, B., and N. Low (Eds). 2001. *Governing for the Environment: Global Problems Ethics and Democracy.* Wiltshire: Palgrave.

Goodin, R. E. 1992. *Green Political Theory.* Cambridge: Polity Press.

Grubb, M. 1995. Seeking fair Weather? Ethics and the International Debate on Climate Change. *International Affairs*, 71 (3): 462–496.

Grubb, M., et al. 1992. Sharing the Burden. In *Confronting Climate Change: Risks, Implications and Responses*, edited by I. Mintzer. Cambridge: Cambridge University Press.

Guha, R. 1989. Radical American Environmentalism and Wilderness Preservation: A Third World Critique. *Environmental Ethics*, 11: 71–83.

Hardin G. (1974). Living on a Lifeboat. *BioScience,* 24: 561–568.

Harmondsworth: Penguin.

Hayek, F. 1960. *The Constitution of Liberty.* London: Routledge and Kegan Paul.

Hayward T. 1998. *Political Theory and Ecological Values*. Cambridge: Polity Press.

1996. Universal Consideration as a Deontological Principle: A Critique of Birch. *Environmental Ethics*, 18: 54–63.

Heyd, T. 2003. The Case for Environmental Morality. *Environmental Ethics*, 25: 4–24.

Holden, B. 1996. Introduction. In *The Ethical Dimensions of Global Change*, edited by B. Holden, 3–7. London: Macmillan Press.

Hurrell, A. 2001. Global Inequity and International Institutions. *Metaphilosophy* 32 (1/ 2): 34–57.

Intergovernmental Panel on Climate Change (IPCC). 1996. *Climate Change 1995: Economic and Social Dimensions of Climate Change, Contribution of Working Group III to the Second Assessment Report of the IPCC.* Cambridge: Cambridge University Press.

Intergovernmental Panel on Climate Change (IPCC). 2001. *Climate Change 2001: Impacts, Adaptation and Vulnerability.* Cambridge: Cambridge University Press.

Intergovernmental Panel on Climate Change (IPCC). 2007. *Climate Change 2007 – Impacts, Adaptation and Vulnerability: Working Group 1I Contribution to the Fourth Assessment Report of the IPCC (Climate Change 2007).* Cambridge: Cambridge University Press.

Jamieson, D. 1992. Ethics, Public Policy and Global Warming. *Science, Technology and Human Values*, 17 (2): 139–153.

Kant, I. 1963. The Duties to Animal Spirits. In Louis Infield (trans.), *Lectures on Ethics*. New York: Harper and Row.

Kelly, P. 1990. The Need for Eco-Justice. *Fletcher Forum of World Affairs* 14: 327–331.

Kymlicka, W. 2002. *Contemporary Political Philosophy: An Introduction* (2nd edn). Oxford: Oxford University Press.

Kutting, G. 2004. Environmental justice. *Global Environmental Politics* (4) 1: 115–121.

Langhelle, O. 1999. Sustainable Development: Exploring the Ethics of Our Common Future. *International Political Science Review* 20 (2): 129–149.

Langhelle, O. 2000. Sustainable Development and Social justice: expanding the Rawlsian Framework of Global Justice. *Environmental Values* 9 (3): 295–324.

Leftwich, A. 1984. *What is Politics? The activity and its study.*

Oxford: Blackwell Books.

Leopold, A. 1949. *A Sand County Almanac.* Oxford: Oxford University Press.

Low, N., and B. Gleeson (Eds). 2001. The Challenge of Ethical Environmental Governance. In *Governing for the Environment: Global Problems, Ethics and Democracy*, 1–13. Wiltshire: Palgrave.

Lucas, G. R., and T. W. Ogletree (Eds). 1976. *Lifeboat Ethics: The Moral Dilemmas of World Hunger.* New York: Harper & Row.

MacIntyre, A. 1981. *After Virtue: A study in Moral Theory.* London: Duckworth Publishers.

Martínez-Alier, J. 2002. *Environmentalism of the Poor: A study of ecological conflicts and valuation.* Cheltenham: Edward Elgar.

McCarthy, J., and S. Prudham. 2004. Neoliberal nature and the nature of neoliberalism. *Geoforum* 35 (1): 275–283.

Meadows, D. H., D. L. Meadows, and J. Randers. 1972. *The Limits to Growth.* London: Pan.

Mies, M., and V. Shiva. 1993. *Ecofeminism.* Halifax Nova Scotia: Fenwood Publications.

Mill, J. S. 1893. *The Utilitarians*, reprinted 1973. New York: Dent.

Mintzer, I., and J. A. Leonard. 1994. *Negotiating Climate Change: the Inside Story of the Rio Convention.* Cambridge: Cambridge University Press.

Mol, A. P. (Ed.). 2000. *Ecological Modernization around the world: perspectives and critical debates.* Ilford: Frank Cass.

Müller-Rommel, F. 2002. The Lifespan and the Political Performance of Green Parties in Western Europe. *Environmental Politics*, 11 (1): 1–16.

Næss, A. 1989. *Ecology, Community, Lifestyle*, translated and edited by D. Rothenberg. Cambridge: Cambridge University Press.

Niemeyer, S. 2004. Deliberation in the Wilderness: Displacing Symbolic Politics. *Environmental Politics*, 13 (2), 347–372.

ESSAYS

Nordhaus, W. 1994. *Managing the Global Commons*. Cambridge, MA: MIT Press.

Nozick, R. 1974. *Anarchy, State, and Utopia*. New York: Basic Books.

Okereke, C. 2007. Global Justice and Neoliberal Environmental Governance: London: Routledge.

2006. Global Environmental Sustainability: Intragenerational Equity and Conceptions of Justice in Multilateral Environmental Regimes. *Geoforum* 37: 725–738.

O'Neill, J. 1992. 'The Varieties of Intrinsic Value', *Monist* 75: 119–137.

1993. *Ecology, Policy and Politics*. London: Routledge.

1997. *Ecology, Policy and Politics: Human Well-being and the Natural World*. London: Routledge

Page, A. E. 2006. *Climate Change, Justice and Future Generations*. Cheltenham: Edward Egar.

2007. Intergenerational justice of what: Welfare, resources or capabilities? *Environmental Politics*, 16 (3): 453–469.

Palmer, M., and V. Finlay. 2003. *Faith in Conservation: New approaches to religion and the Environment*. Washington: IBRD/World Bank Publications.

Pardo, A. 1975. *The Common heritage: Selected Papers on Ocean and the world order*, 1,967–74. Valletta, Malta: University of Malta Press.

Paterson, M. 1996. International Justice and Global Warming. In *The Ethical Dimensions of Global Change*, edited by B. Holden, 181–201. London: Macmillan Press.

Paterson, M. 2001. Principles of justice in the context of global climate change. In *International Relations and Global Climate Change*, edited by U. Luterbacher and D. Sprinz. Cambridge: MIT Press.

Pepper, D. 1984. *The Roots of Modern Environmentalism*. London: Routledge.

Plumwood, V. 1993. *Feminism and the Mastery of Nature*. London: Routledge.

Plumwood, V. 2002. *Environmental Culture: the Ecological Crisis of Reason*. London: Routledge.

Ramakrrishna, K. 1990a. North South issues, Common Heritage of mankind and Global Climate Change. *Millennium Journal of International Studies*, (19) 3: 429–445.

1990b Third Word Countries in the Policy Response to Global Climate Change. In *Global Warming: The Greenpeace Report*, edited by J. Leggett, 421–437. Oxford: Oxford University Press.

Rawls, J. 1971. *A Theory of Justice*. Oxford: Oxford University Press.

Rolston, H. 1989. *Philosophy Gone Wild*. New York: Prometheus Books.

Routley, R., and V. Routley. 1980. Human Chauvinism and Environmental Ethics. In *Environmental Philosophy*, edited by D. M. A. Mannison and Routley,

156

R. 96–198. Canberra: Australian National University Research School of Social Sciences.

Sachs, W. 1999. *Planet Dialectics: Exploration in Environment and Development*. London: Zed Books.

Sagoff, M. 1995. Can Environmentalists be Liberals? In *Environmental Ethics: Oxford Readings in Philosophy*, edited by R. Elliot. Oxford: Oxford University Press.

Sanger, C. 1986. *Ordering the Oceans: The Making of the Law of the Sea*. London: Zed Books.

Schlosberg, D. 2005. Environmental and Ecological Justice: Theory and Practice in the United States. In *The State and the Global Ecological Crisis*, edited by J. Barry and R. Eckersley. Cambridge, MA: MIT Press.

Shue, H. 1995. Ethics, the Environment and the Changing International Order, *International Affairs* 71 (3): 453–461.

Shue, H. 1993. Subsistence emissions and luxury emission. *Law and Policy* 15 (1): 39–59.

Singer, P. 1975. *Animal Liberation*. New York: Random House.

1993. *Practical Ethics* (2nd edn). Cambridge: Cambridge University Press.

Singer, S. F. 1992. Earth Summit Will Shackle the Planet, Not Save it. *The Wall Street Journal*, 19 February 1992.

Stern, N. 2007. *The Economics of Climate Change: The Stern Review*. Cambridge: Cambridge Press.

The Founex Report on Development and Environment. 1971. Geneva: Southcentre Publications.

Thompson, J. 2000. Environment as a Cultural Heritage. *Environmental Values* 22, 241–258.

Vanderheiden, S. 2005. Missing the Forest for the Trees: Justice and Environmental Economics. *Critical Review of International Social and Political Philosophy*. 8 (1): 51–69.

Vogler J. 2000. *The Global Commons Environmental and Technological Governance* (2nd edn). Chichester: John Wiley Press.

Weale, A. 1992. *The New Politics of Pollution*. Manchester: Manchester University Press.

White, L. 1967. The Historical Roots of Our Ecological Crisis. *Science*, 55: 1,203—1,207; reprinted in Schmidtz and Willott 2002.

World Commission on the Environment and Development (WCED). 1987. *Our Common Future*. Oxford: Oxford University Press.

Young, O. 1989. *International Cooperation: Building Regimes for Natural Resources and the Environment*. Ithaca, New York: Cornwell University Press.

A–Z Glossary

By Chukwumerije Okereke

A

Acid Rain

A terms that is generally used to cover the deposition of acidic precipitation in the form of rain, hail, snow, fog or solid particles on the Earth's surface. Acid rain results when air pollutants that are released into the atmosphere from the use of motor vehicles and other industrial processes, especially the burning of fossil fuels (coal or petroleum) in power stations, react with tiny droplets of water vapour in the atmosphere to form dilute mineral acids which are then deposited back in the form of precipitation onto the Earth's surface. There are many different types of air pollutants that can undergo this process to form acid rain, but the most popular are particles of sulphur dioxide and nitrogen oxide. When these substances react with water vapour in the atmosphere they form sulphuric acid, nitric acid, and ammonium salts which confer a high degree of acidity on the precipitation. Acidity is measured using what is normally called a pH scale. A pH scale ranges from 0 to 14, with 0 being the most acidic while 14 is the most alkaline. A substance with a pH value of seven is said to be neutral—it is neither acidic nor alkaline. Normal precipitation is slightly acidic (between the pH values of five and six) because water vapour ordinarily reacts with some naturally occurring oxides in the atmosphere. However, when the atmosphere becomes heavily polluted with these nitrous and sulphuric oxides, the pH value of precipitation could be reduced to four or even as low as two on some occasions. There is high variability in the distance between source of pollution and the point of deposition. While some precipitations are deposited a few miles from the source of pollution, others are carried high into the clouds and travel with the winds for hundreds of miles before falling back on the Earth's surface as dust, rain, mist or snow. However, acid rain is generally a regional rather than a global phenomenon, although some scientists have suggested that a small portion of the acid rain in Europe arises from polluting activities in the USA. Acid rain can cause all sorts of problems both to human lives and the natural environment. It results in the rapid corrosion of building roofs, motor vehicles and public statues, as well as plumbing pipes and cables. Acid rain may also lead to the deaths of living organisms in lakes and freshwater environments. Some lakes in Scandinavia have been reported to be completely devoid of life due to acid rain. Acid rain can also lead to the destruction of forests as trees either find it difficult to grow at their normal rates or even die altogether as the soil becomes increasingly acidic. Acid rain was one of the main environmental problems that were highlighted during the first UN Conference on the Human Environment at Stockholm in 1972. The major complaints came from Norway and Sweden, which argued that pollution arising from the United Kingdom was causing the destruction of several of their lakes and rivers. However, this problem had long been evident in Canada and the USA, which passed a Clean Air Act in 1963 as part of the measures to deal with the problem. Acid rain is now one of the major environmental problems in China where about one-third of the landmass (696 cities) is said to be severely affected, with soil quality, food safety and human health all being threatened.

Adaptation

Adaptation generally refers to the process of modifications or adjustments in behaviour, aspects of operations or rules in order to respond to changes in the external environment. In the context of environmental politics the term is used mostly in climate change discussions and refers to the changes that societies are supposed to make in order to respond to the negative impacts of unavoidable climate change. It is also used in discussions that focus on the tendency as well as the limit of the Earth's natural system and living organisms to adjust in ways that can accommodate the process of (natural and human-induced) environmental change without resulting in catastrophic consequences. When the issue of global climate change became a major political topic in the late 1980s and early 1990s, the main focus was on mitigation—that is, how to reduce carbon dioxide emissions in the atmosphere in order to prevent the incidence of human-induced climate change. The focus on mitigation is clearly reflected in the wordings of the UN Framework Convention on Climate Change (UNFCCC) which has as its objective the need to stabilize greenhouse gas concentration in the atmosphere at a level that would prevent dangerous interference with the climate system, although the word adaptation was subsequently mentioned in a number of places in the convention's text. Adaptation received less attention in the early years mainly because parties to the convention requested more certainty on the effects of climate change on different natural and social systems, and on their vulnerability to them. In 2001 the **Intergovernmental Panel on Climate Change (IPCC)**, which is an international scientific body responsible for providing policy-makers with unbiased scientific information on climate change, reported that it was reasonably certain that climate change was already happening, with considerable negative effects on several ecosystems and human settlements around the world. The major effects were said to be the general reduction in crop yield in most tropical regions, increased flooding of human settlements, water scarcity and the increase in water- and vector-borne diseases such as malaria due to drastic changes in precipitation patterns. Those identified to be most vulnerable to the negative impact of unavoidable climate change were the Third World countries, especially the Least Developed Countries (LDCs) and the Small Island Developing States (SIDS). The panel called for attention to be given to adaptation as well as mitigation in the global effort to tackle climate change. Since then a huge number of studies from both public and private organizations have been conducted to ascertain the scale of impact and the vulnerabilities of various ecosystems and human communities to climate change. Most of the results show that the developing countries which contribute little to climate change are the ones that are the most vulnerable to its negative consequences. Subsequently the developing countries have been pushing for adaptation to be accorded a priority in the scheme of things in the UNFCC. At the 10th Conference of Parties (COP) meeting in Buenos Aires, Argentina, in 2004 a plan of action on adaptation and response measures was adopted as part of the response to this campaign. There have also been other programmes, including the establishment of a special needs fund for the LDCs (the LDC Fund) and a five-year programme of work on impacts, vulnerability and adaptation, all designed to help the developing countries to respond to the impacts of climate change.

Agenda 21

One of the five main documents adopted at the UN Conference on Environment and Development (UNCED) in Rio de Janeiro, Brazil, in 1992, Agenda 21 is a comprehensive plan of action that details the steps that need to be taken by governments, businesses and civil society in order to redirect the Earth towards a more secure, more equitable and more sustainable pathway. The other four documents were the Rio Declaration on the Environment and Development, which defines the rights and responsibilities of nations as they strive to achieve economic development and environmental sustainability; a statement of principles to guide the management of the world's forests; and two international conventions—one on climate change and the other on **biodiversity**. Agenda 21 reflects a global consensus and political commitments regarding the need, and how to implement the fundamental changes required to achieve global **sustainable development**, as was called for at the Rio **Earth Summit**. The document explains that overconsumption and population are the primary driving forces for environmental change and urges governments to take practical steps to deal with issues of overconsumption as well as global poverty and inequity. It calls on governments to develop broad-based partnerships with businesses, **non-governmental organizations (NGOs)** and citizens' groups to work out how best to establish institutions that would help the world to make the transition to sustainable forms of development and lifestyles. Agenda 21 is a significant contribution to international co-operation for sustainable development, in terms of both its content and, perhaps even more so, of the process by which the document was developed. In the preparations for the Earth Summit, the UN set up a panel of experts which toured different parts of the world and held open deliberative meetings with thousands of people from businesses, NGOs, women's groups, indigenous groups and government officials. It was in the process of these meetings that the document was developed and refined. The main significance of the process was that it opened up the international negotiation system as never before and established an unprecedented level of broad public participation in the formulation of national and international sustainable development policies. The procedure through which Agenda 21 was developed might be called a snapshot of what global deliberative democracy could possibly look like in the future. But well over two decades after the adoption of this historic document there are indications that the massive expectations of the world's people of their governments are not being realized.

Antarctica

Antarctica is one of the seven continents of the Earth. It is the fifth largest continent, with an area of about 14m. sq. km (about twice the size of Australia). The continent is centred on the South Pole and surrounded by the Southern Ocean. It is the Earth's southernmost continent. Antarctica is the coldest continent, with about 97% of its surface covered by ice averaging about 1.6 km in thickness. Antarctica is equally the windiest and driest continent, with a very low amount of precipitation, much of which occurs at the coastal margins. There are seven countries that lay claim to different portions of Antarctica. The list of the claimant countries and the dates of their claims are: United Kingdom (1908); New Zealand (1923); France (1924); Norway (1929); Australia (1933); Chile (1940); and Argentina (1943). There are, however, no permanent human residents in Antarctica. Indeed the continent remains the least well known and is thought to contain a variety of valuable non-renewable mineral resources. Antarctica is also seen as having a very strong relationship with the global ecosystem and climate system because it not only acts as a sanctuary for some significant species of flora and fauna but also has a cooling effect, which affects the global atmosphere, wind and ocean circulation. For these reasons there has been much political, economic and environmental interest in Antarctica. On 1 December 1959 a treaty designed to protect Antarctica was signed by 12 states in Washington, DC, USA. The original signatories were the 12 states active in Antarctica during the International Geophysical Year of 1957/58. These countries include Argentina, Australia, Belgium, Chile, France, Japan, India, New Zealand, Norway, South Africa, the (former) USSR, the United Kingdom and the USA. These original signatories have consultative status with voting power, but all members of the UN are permitted to accede to the convention. To qualify to become a consultative party, a country has to demonstrate scientific interest and presence in Antarctica. There are now 28 consultative members and a total of 46 accessions to the treaty. The treaty, which entered into force on 23 June 1961, stipulates that Antarctica should be used exclusively for peaceful purposes and forbids all types of military activity and mining in the continent. The treaty also prohibits nuclear explosions and the disposal of radioactive waste and encourages international scientific co-operation, including the exchange of research plans and personnel in order to enhance understanding of the continent. The treaty does not annul pre-existing claims of sovereignty but provides that no activities will enhance or diminish previously asserted positions with respect to territorial claims and that no new or enlarged claims can be made. Both the the USA and Russia have indicated that they do not recognize the territorial claims of the other countries, but at the same time that they reserve the right to make claims in the future if they consider it necessary. Since then other measures have been adopted by states to enhance the protection of Antartica and its natural environment. These include the Agreed Measures for the Conservation of the Antartica Fauna and Flora (1964); the Convention for the Conservation of Antartica Seals (1972); the Convention for the Conservation of the Antarctic Marine Resources (1980); and the Convention on the Regulation of Antarctica Mineral Resource Activities (1988). During the elaboration of the Convention for the Regulation of Antarctica Mineral Activities the USA and its allies made a proposal that would have allowed limited mining on the continent. But the proposal was vehemently opposed by France and Spain. On 4 October 1991 governments eventually signed the Protocol on Environmental Protection to the Antarctica Treaty which bans mineral and petroleum exploration in Antarctica for at least 50 years. Some developing countries complain that the provision for consulative status on the basis of scientific activities is exclusionary and that membership should be open to all countries of the world since Antarctica is a common heritage of all mankind.

Anthropocentrism

Anthropocentrism refers to the idea that humans are the ultimate intention of creation and the final purpose of the universe. It is the notion that only human beings have moral worth, and that all other things are valuable only insofar as, and to the extent that, they are deemed to be so by human beings. Anthropocentrism is often contrasted with biocentrism which is the notion that other beings and natural objects apart from humans equally have intrinsic value and therefore are worthy of moral consideration from human beings. Anthropocentrism is regarded as the dominant Western philosophy. Some trace the roots of the Western anthropocentric worldview to the Bible, citing instances in the book of Genesis where it is reported that God commanded man to be fruitful and to have dominion over all the other creatures on the Earth. There is, however, evidence that all through the ages the notion that humans are the central element of the universe has always been a major form of human consciousness although the notion that humans should have a deep appreciation and more than instrumental relationship with non-human nature is equally a significant canon in much ancient philosophical thought and many popular world religions. The question of what should be the moral relationship between human and non-human beings has been around for a long time but it was only more recently that some began to mount a serious challenge against the anthropocentric worldview. Some environmentalists argue that the anthropocentric worldview is at the root of the current widespread **environmental degradation** and drastic changes in the global environment. Others go so far as to attempt to establish a link between anthropocentrism and other social vices, such as racism, chauvinism and the oppression of women. They argue that all of these social vices relate to anthropocentrism in the sense that they are all united by the same idea of discrimination and domination by a privileged class. Critics make a powerful statement against anthropocentrism but they generally find it more difficult to establish why non-human life forms and natural phenomena like ants, trees, bacteria, rivers and microbes should be accorded the same moral status as human beings.

B

Basel Convention

This is shorthand for the Basel Convention on the Control of Transboundary Movements of Hazardous Wastes and their Disposal negotiated in Basel, Switzerland, in 1989. The convention was the product of a world-wide outcry against the indiscriminate dumping of hazardous wastes on developing nations and Eastern European countries by waste brokers from the industrialized states in the early 1980s. The sharp increase in the dumping of hazardous wastes on the developing countries of the world is usually attributed to the tightening of environmental regulations in industrialized countries which led to a dramatic rise in the cost of hazardous waste disposal in these countries. The other factor is the increase in socio-economic globalization which offers waste brokers from the industrialized countries several routes to dispose of their wastes at costs that are usually far cheaper than they would have paid if they had disposed of them in the countries of origin. Developing countries, however, complained that the rapid shipments of hazardous wastes into them were posing all sorts of health and environmental risks given that they do not have adequate facilities to dispose of them in an environmentally sound manner. They used terms such as 'waste colonialism' and 'trash imperialism' to describe the continuous shipment of hazardous wastes to them despite their opposition to and inability to deal with such wastes. It was as a result of this outcry that the Governing Council of the **UN Environmental Programme (UNEP)**, through Decision 14/30 of June 1987, authorized the executive director to set up a working group with the task of articulating a comprehensive text on all aspects of transboundary movements of hazardous wastes and their disposal. The Governing Council in the same decision also mandated the executive director of UNEP to convene a diplomatic conference in 1989, at the latest, for the purpose of adopting a global convention on the control of transboundary movements of hazardous wastes. The result was a series of diplomatic negotiations that culminated in the development of the Basel Convention on the Control of Transboundary Movements of Hazardous Wastes and their Disposal in March 1989. The highlight of the Basel Convention was the establishment of Prior Informed Consent (PIC), which made it mandatory that waste brokers must first obtain the express consent of countries of destination before shipping wastes to such countries. The convention also provided for the establishment of focal points and competent authorities by the parties to provide clearance to those seeking to import wastes into these countries. In September 1995, at the third meeting of the COP, parties agreed to ban altogether shipments of hazardous wastes from OECD countries to non-OECD countries. In December 1999 the Basel Protocol on Liability and Compensation for Damage Resulting from Transboundary Movement of Hazardous Wastes and their Disposal was negotiated to provide for effective and speedy compensation for those that suffer injury as a result of the transboundary movements of wastes and their disposal. The Basel Convention is one of the most ratified global environmental treaties. The convention entered into force on 5 May 1992 with the deposition of the 20th instrument of ratification. But there are now about 165 parties to the convention, including the European Union. Afghanistan, Haiti and the USA have all signed but not ratified the convention. More recently, the focus of the Basel Convention has shifted from just controlling the movements of hazardous wastes across countries to the environmentally sound management of wastes and the minimization of waste production generally.

Bhopal

Bhopal is the capital city of the state of Madhya Pradesh in the central part of India. This city, which has a population of about 1.5m., was relatively unknown until 3 December 1984 when it became one of the best-known instances of environmental disaster in the world. Tragedy struck in the early hours of the morning on 3 December when a holding tank with about 45 tons of methyl isocyanate gas (MIC) at a plant owned by Union Carbide India Ltd (UCIL), a subsidiary of Union Carbide International (UCI), headquartered in the USA, overheated and released this highly poisonous mixture into the densely populated areas immediately around the location of the plant. Soon after, several hundred people choked to death and many more were trampled to death as residents attempted to flee the city in panic. Official figures estimate that about 3,500 people died as a result of the incident and that the health of over 200,000 people has been significantly affected by the incident. Others sources suggest that well over 20,000 people died either immediately or soon afterwards as a result of complications arising from inhalation of the poisonous gas. After the incident the Indian Government in March 1985 passed a Bhopal Gas Leak Act which allows the Government to act as a litigant on behalf of those who have been affected by the gas leak and subsequently brought a lawsuit against Union Carbide in the US courts, claiming US $3,000m. in damages. UCI, however, objected to the case being heard in a US court of law. The company argued that the Bhopal plant was owned and operated by UCIL and that UCIL was a separate entity from UCI. The US court upheld this objection and had the case transferred to an Indian court. After a protracted legal battle the parties reached an out-of-court settlement on the advice of the Supreme Court of India in 1989, with Union Carbide agreeing to pay $470m. in compensation for damages. There have subsequently been several other cases both in India and the USA instituted by activist groups against the settlement, but none of these have been upheld by the court. In 2006 a New York court upheld the dismissal of the remaining claims, insisting that all cases relating to the incident should be instituted and pursued in India. However, the Indian Supreme Court had in July 2004, in a separate lawsuit, ruled that the $470m. out-of-court settlement was conclusive but ordered the Government to release any remaining settlement funds and ensure that the monies reached the victims without further delay.

Biodiversity

Biodiversity or biological diversity describes the variety of all forms of life on Earth, including their behaviours and interactions as well as the ecological complexes of which they are part. The term is often used in three different but related ways. The first is with reference to the diversity among living organisms on Earth, including plants, animals and micro-organisms. The second is with reference to the different types of ecosystems that obtain around the world. This might be with respect to broad categories such as the aquatic or terrestrial environment, or to more precise descriptions such as wetlands, mangroves, mountains, rivers, lakes, forests, deep sea-bed, deserts, etc. At a third level, biodiversity is used to describe the varieties and differences in individual species of flora and fauna. Some also use the term to refer to the totality of the genes and chromosomes that constitute the building blocks of life and confer uniqueness to various species by reason of their different arrangements in the cells of living organisms. Currently scientists have identified up to 1.8m. species on Earth. Most of these species are small creatures like insects and tiny plants such as algae and fungi. There are huge variations in the estimates of the number of all the different types of life on Earth. While some put the figure at about 13m., others suggest that the number could be up to 80m. The number and the richness of biodiversity are functions of natural systems as well as human activities. Even if humans were not to interfere with nature in any way, it is believed that natural variations in weather patterns would still lead to progressive changes in the biotic make-up of any given ecosystem over time. However, there is equally a widely shared concern that human activities over the past few decades, especially those connected with industrial development, have resulted in rapid depreciation in the variety and number of the different forms of life on Earth. Scientists suggest that human activity has forced over 800 species into extinction in the last 500 years. They estimate that human activity is causing an increase of between 10 and 1,000 in the rate of species extinction compared to the 'natural' rate. Currently over 15,000 species of plants and animals are considered to be under the threat of extinction. There is a feeling that the rapid extinction of species would affect not only the quality of the world's ecosystems but ultimately the quality of human lives. This concern is the main reason why the world's leaders, at the **Earth Summit** in Rio de Janeiro, Brazil, in 1992 negotiated a UN Convention on Biodiversity. Governments have also set aside 22 May of every year to mark the International Day of Biodiversity.

Biosafety Protocol

A protocol to the 1992 UN Convention on Biological Diversity (CBD) which has the objective of ensuring the safe transfer, handling and use of living modified organisms (LMOs)—also known as **genetically modified organisms (GMOs)**—in order to avoid potential adverse effects on the conservation and sustainable use of native **biodiversity** and its components. The Biosafety Protocol was negotiated because of some expressed fears that the uncontrolled and unregulated transboundary movements of GMOs could result in dangerous interference with the natural biodiversity of countries to which they are exported. Based on this concern, the parties to the UN Convention on Biological Diversity in November 1995 established an Ad Hoc Working Group on Biosafety to develop a draft text that might guide parties in the control of transboundary movements of GMOs. After about four years of work, the Ad Hoc Working Committee presented a draft text to the Conference of Parties (COP) in its first extraordinary meeting in Cartagena, Colombia, in 1999. The text was debated but a final decision could not be reached. The text was, however, adopted in a resumed session of the COP in Montréal, Canada, in January 2000. The main highlight of the protocol is the establishment of the Advanced Informed Agreement (AIA) rule, which requires that the consent of designated authorities in the country of import must be sought and obtained before a shipment of modified living organisms can be undertaken. The other is the establishment of a Biosafety Clearing House to facilitate the exchange of information on LMOs and to assist countries in implementing the protocol. About 133 countries have ratified the protocol, but three of the main countries that grow genetically modified crops—Canada, the USA and Argentina—are still opposed to the convention. One of the main reasons given by these countries is that the protocol constitutes undue restriction on international trade. Other major challenges that have been encountered by parties to the convention include the difficulty in agreeing to the exact nature and level of detail of information and documentation that should accompany shipments of LMOs and matters relating to capacity-building and technology transfer from the developed to the developing countries.

Bogotá Declaration

A political statement made by eight equatorial countries on 3 December 1976 in Bogotá, Colombia, claiming sovereignty over aspects of the geostationary orbit (GSO). The list of the countries includes Brazil, Colombia, the Democratic Republic of the Congo, Ecuador, Indonesia, Kenya and Uganda. The GSO is a highly unique orbit, which circulates the Earth along the equator at an altitude of about 35,871 km above the Earth. A satellite launched into this orbit will travel at exactly the same speed as the speed of the Earth's rotation and in the same direction. Because of this, satellites launched into this orbit appear to be stationary in the sky when viewed from the Earth, hence the name geostationary orbit. GSO is an extremely valuable resource because satellites launched into this orbit provide a highly reliable means of maintaining steady contact with earth-bound transmitters as well as extensive coverage of the Earth. Nearly the entire globe could be covered by three satellites placed in the GSO. The main argument of these equatorial countries was that the GSO is not in that part of outer space which is designated as the **common heritage of mankind** in international law. Rather, the orbit should be considered as a natural overhead extension of their national territories and therefore subject to the claims of national sovereignty. In making this claim, the countries relied on two key arguments. Firstly, they argued that the GSO has a fixed location which can be mapped in relation to exact physical locations on the Earth's surface. They then claimed that insofar as the GSO is located directly above the Equator, it should be considered as part of the territory of the equatorial countries. Secondly, they argued that conferring ownership of the GSO on the equatorial countries could be seen as a means of promoting global justice and equity. They pointed out that the GSO is a limited natural resource. Not only could satellites be placed in permanent positions in the GSO, it is also possible to reach a point of saturation when no additional placement of satellite is possible. The declaration document argued that it is not fair that the developed countries, which have superior technological power, should be allowed to have a monopoly on the use of the orbit without sharing the benefits with the developing countries. But besides immediate attention following the statement, the Bogotá Declaration has not had much influence on the development of the international law of outer space. However, there is no guarantee that the issues raised in the declaration will not be revived in the future if the availability of the GSO becomes an issue of concern.

Bonn Convention

The Bonn Convention on the Conservation of Migratory species of Wild Animals (CMS) is an intergovernmental treaty developed in 1979 in Bonn, (Federal Republic of) Germany, to protect terrestrial, marine and avian migratory species considered to be endangered through all or part of their range. The range is defined as all the areas of land or water that a migratory species inhabits, stays temporarily in, crosses or overflies at any time on its normal migratory route. The convention was developed under the auspices of the **United Nations Environ-mental Programme** and currently has over 100 states as parties to the treaty. The Convention's Appendix I contains a list of about 123 migratory species and subspecies which are considered to be threatened with extinction. The treaty imposes strict conservation responsibilities upon member states whose territories provide the habitat or corridors for these endangered migratory species. There is also Appendix II, which contains a list of about 200 migratory species. Animals in Appendix II are considered to have unfavourable conservation status and would benefit from either a formal conservation treaty or a memorandum of understanding (MOU) from the range states. In other words, the convention serves as global framework under which many other agreements and bilateral and regional treaties as well as MOUs might be developed. So far about six other agreements and 11 MOUs have been signed under this framework convention. Some of the more notable ones include the African-Eurasian Water Bird Agreement and the Agreements on the Conservation of Populations of European Bats (EUROBATS), which aims to conserve 45 out of the approximately 1,100 known species of European bats.

Brundtland Report – *see* **World Commission on Environment and Development Report (WCED)**

C

Carbon Disclosure Project (CDP)

An initiative launched in London on 4 December 2000 at 10 Downing Street as part of the efforts by a group of institutional investors to encourage major corporations around the world to provide regular and ample information about their greenhouse gas emissions and carbon-management activities. The Carbon Disclosure Project (CDP) serves as a form of secretariat for institutional investors seeking information on the risks and opportunities presented by climate change and particularly the greenhouse gas emission profiles of the world's largest companies. The assumption is that this information will feed into their investment decisions. The main activity of the secretariat, a branch of which is located in New York, USA, is to send out questionnaires probing into the greenhouse gas emissions and general carbon-management strategies of corporations to big businesses around the world once a year. The results of these questionnaires are collated and analysed by environmental consultancy firms and subsequently made public through the CDP website. This then provides a major resource not just for institutional investors but also for governments and private individuals wishing to track the emission profiles or carbon-management activities of companies. So far there have been four reports: CDP1 (2003)–CDP4 (2006). Within this period there has been a remarkable increase in both the number of institutional investors that sign the questionnaires and the number of companies that are pooled. For example, while the CPD1 information request was signed by just 35 institutional investors and sent to the FT500 largest companies, the CDP4 questionnaires were signed by 225 institutional investors with assets worth more than US $31,000,000m. and sent to 2,180 companies. The latest information request, CDP5, was signed by 280 institutional investors with assets worth more than $41,000,000m. and sent to 2,400 companies. The results and analysis of this are due to be published in the last quarter of 2007. The CDP website has ultimately become the largest repository of corporate greenhouse gas emissions data in the world. The project has had tremendous influence, especially in terms of forcing companies to make their emission profiles public, but also to take action on climate change. The information collated and published under the CDP also provides a powerful resource tool for academics and policy-makers.

Carbon Tax

As discussions on how to tackle the challenge of global warming become more intense and widespread, an increasing number of people are beginning to advocate the imposition of direct taxes on businesses and private individuals to reflect their emissions of greenhouse gases. All across the world, there are already several types of energy taxes that are mostly weighted against producers. There is equally a growing number of 'green taxes' that are designed to encourage people to behave in an environmentally benign manner. The idea of asking companies and individuals to pay directly for the emission of climate-warming gases in the form of taxes is a relatively new one, but is none the less gathering momentum. Already the government of the city of Québec in Canada has announced plans to introduce a carbon tax under which companies will pay an extra 0.8 cents on every litre of gas sold. The money raised from this tax will be used to fund green projects such as the installation of **renewable energy** facilities. Similarly, in New Zealand, the Government has announced plans to introduce an extra amount of money on electricity, petrol and gas use as part of the efforts to limit greenhouse gas emissions and generate funds for climate-friendly projects. The European Union has considered the option of imposing a carbon tax among its 27 member states to complement the EU Emissions Trading Scheme (EU ETS) and other national greenhouse emission-reduction policies. While the idea of carbon tax is relatively new, there is every indication that it will become a more popular instrument among the list of other policies that will be used by governments in combating the challenge of climate change.

Chernobyl

Chernobyl (Chornobyl) is a city in Northern Ukraine where the world's worst-ever nuclear disaster occurred in the early hours of 26 April 1986. At exactly 01:23 a.m. local time, one out of the four nuclear reactors in a nuclear plant located about 14 km north west of the city blew the reinforced concrete top from the graphite-moderated channel tube reactor, subsequently releasing a huge amount of radioactive waste into the atmosphere. It has been reported that the accident occurred when some operators at the plant were conducting experiments with the reactor's safety system. Others have suggested that the accident was due to the prevalence of badly maintained equipment and poorly trained operators in the plant. Whatever the cause of the accident, the brute fact is that large quantities of the released radioactive waste travelled several kilometres, reaching many parts of northern Europe and resulting in what some still consider to be Europe's single biggest environmental problem to date. The explosion resulted in the evacuation of the approximately 135,000 residents and millions of head of livestock from the city. Of the 100 firefighters who were called to the accident site, 31 died after just one week due to radiation poisoning, and over 200 subsequently developed serious illness as a result of exposure to radiation. It is estimated that over 600,000 people were exposed to various doses of radiation and stand a considerable chance of developing cancer within their lifetime. They now need to be regularly monitored for the rest of their lives. Already, birth defects among children in Ukraine have increased by over 230% and it is feared that the worst consequences will be seen in generations yet unborn. Cleaning operations have been under way for years, with more than half a million trained people already having been involved. Nevertheless, most of the surrounding villages are likely to remain quarantined for many decades if not centuries. There are, however, about 1,500 mostly elderly men who have returned to live in some of the cleaned-up areas in the city.

CITES

CITES—the Convention on International Trade In Endangered Species of Wild Flora and Fauna and their products— is unique in more ways than one. Firstly, it is one of the very few environmental conventions that were initiated within a broader forum comprising state governments, **non-governmental organizations (NGOs)** and individual scientists. The text of CITES was drafted as a result of a resolution adopted at a meeting of members of the **World Conservation Union (IUCN)** in 1963. The text of the convention was finally agreed at a meeting of representatives of 80 countries in Washington, DC, USA, in March 1973. The convention eventually entered into force on 1 July 1975. It now has over 172 members as parties. CITES is also one of the few environmental conventions that rely solely on the control of trade as a means of achieving conservation objectives. Hence, it is strictly speaking a trade rather than an environmental regime. Nevertheless, CITES has played a significant role in the preservation of endangered species across the world, all the more so because it is one of the few multilateral environmental treaties that may impose sanctions for non-compliance. CITES was negotiated on the logic that the loss of many important species of flora and fauna around the world is not only due to the destruction of their habitats, but also to the lucrative trade in these species and their products, which leads directly to overexploitation. Trade in various plant and animal species and their products is a big business worth billions of dollars per year. This consists of trade in ornamental plants and in live exotic animals which are kept in zoos in foreign countries and as pets in wealthy homes, or else used as educational materials in schools. There is also trade in a vast array of wildlife products, such as fur, skins, timber, leather, wooden musical instruments, medicines, shells and other artefacts. CITES works to control trade in endangered species through the use of a permit system. It has three appendices. Appendix I contains the list of the highly endangered species of flora and fauna and their products. Trade in these are, with some exceptions, generally prohibited. Appendix II contains the list of species which are not necessarily threatened with extinction but in which trade is deemed to require some control to ensure sustainable utilization. Appendix III contains lists of species that are protected in at least one country, usually at the discretion of a single party member. Each country member has a management authority and a scientific authority which work in conjunction to issue permits for the import and export of listed species. In total, about 5,000 species of animals and 28,000 species of plants are protected by CITES. The most notable of these include African elephants and their ivory, rhinos, the giant panda, whales and tigers. The secretariat of CITES is located in Geneva, Switzerland, and administered by the **United Nations Environmental Programme**. The members meet once every two years to review the lists based on scientific evidence and to decide on other matters that might be necessary to strengthen the convention.

Club of Rome

An association formed in 1968 by an Italian industrialist, Aurelio Peccei, and a Scottish scientist, Alexander King, to act as a global think-tank on political, socio-cultural, ecological and technological issues. The Club of Rome has mostly eminent people from different professions and continents as members. The association has as its main objective the desire to generate ideas and initiate critical debates on issues that are considered fundamental to the improvement of global society as a whole. The main strategy through which this aim is pursued is by organizing conferences and commissioning reports. The Club of Rome prides itself on taking a more long-term perspective on issues than government would normally do and also on taking a multidisciplinary approach to problem-solving. The main areas of interest have been in demography and urban population, inflation and world economic development, resource use and ecological sustainability, information technology and global justice. The Club of Rome was part of the movement that recommended a New International Economic Order as a means of lessening the pervasive inequality between the developed and the developing countries of the world in the early 1970s. However, by far the most influential report commissioned by the organization is **Limits to Growth** (1972)—which claimed that under the then business approach, the world would exhaust its natural resources in the next 100 years and would be subsequently plunged into great depths of economic depression and socio-political instability. This report has been translated into more than 30 languages and has sold over 30m. copies. The report was well received by many ecologists but was heavily criticized by economists who branded it a doomsday report. Nevertheless, it played a very significant role in increasing environmental awareness across the globe and in pushing the issue of resource limits and sustainability onto the international agenda. The co-founder and former president of the club, Alexander King, died on 28 February 2007 at the age of 98.

Collective Action Problem

This term is used to describe the specific challenges associated with the inherent tendency of people who should otherwise have long-term interests in preserving a commonly-owned resource to actually overexploit such resources for individual short-term benefits, even with the knowledge that similar behaviours by other legal users would ultimately result in the degradation of the commonly-owned resource. The concept was popularized by two scholars in the mid-1960s. The first was Mancur Olson, Jr, who published a much-cited work entitled *The Logic of Collective Action* in 1965. The other was Garrett Hardin, whose work, *The Tragedy of Commons* (1968), perhaps did more than any other to entrench the concept in the vocabulary of environmental politics. Both scholars argued against the received wisdom of the time, which suggested that people who have interests in the use of a commonly-owned resource would naturally act collectively to preserve such a common. They argued rather that there is an inherent tendency for rational egoistic legal users to 'free-ride', that is, to abuse the resources for their short-term gain at the expense of the collective interest of the group. The problem, they argued, is that while everyone knows that the overall long-term interest of the group is best served by conserving the resource, there is also a persistent feeling that if they don't overuse the resource, some other person will. Moreover, there is also the feeling within each single user that overuse of the resource would not do much harm to it. They argued that these two instincts combine to cause most users to overexploit, leading eventually to complete degradation of the commons. Such was the assumed fate of all commons that Garrett Hardin coined the term 'the tragedy of commons' to describe this counterintuitive logic. There are many people who describe most of the world's environmental challenges in terms of collective action problems. They suggest that environmental problems arise because while many people mentally appreciate the need to act in environmentally benign ways, there is also the inherent tendency to abuse resources for short-term economic gains.

Common but Differentiated Responsibility (CDR)

A concept in international environmental law which basically provides that developing countries might need to take lighter responsibilities than the developed countries in the collective effort to deal with global environmental problems. Common but Differentiated Responsibility (CDR) is generally regarded as an equity principle through which it is acknowledged that, although all the countries of the world share a common need to respond to the threats of global **environmental degradation** and drastic changes in the global environment, developed countries would need to bear greater responsibility since they are on balance more responsible for causing the problem. The concept also serves as a means of recognizing that certain environmental standards agreed for the developed countries might not be suitable for the developing countries, given their (the developing countries') special economic and developmental needs. At the same time, the CDR embodies a recognition of the wide disparities in the economic, scientific and technological capabilities of states and the logic that governments should make contributions that are commensurate with their capabilities in the collective search for solutions to global environmental challenges.

Common Heritage of Mankind (CHM)

A concept coined by the former Permanent Representative of Malta to the UN, Ambassador Arvid Pardo, to describe the deep sea-bed beyond the jurisdiction of states and the mineral resources said to be located in this area. In a moving speech before the UN General Assembly in December 1967, Ambassador Pardo urged the UN to quickly elaborate a global regime that would provide the framework for the protection of the deep sea-bed and the minerals therein from being used for military purposes or being exploited by a handful of industrialized countries that have the technical capabilities to do so. In contrast to the concept of *res nullis* in which the resources of the oceans and the high seas are regarded as belonging to no one, Pardo argued that these areas should be viewed as commonly inherited and owned by all mankind. In line with this proposed concept of ownership, he proposed that the resources from these areas should be used to offset the economic inequity between the developed and the developing countries. The argument linking the common heritage of mankind concept with global equity proved so popular that it became the mantra of the group of developing countries (**G77**) throughout the period of the negotiation of the third UN Law of the Sea (UNCLOS III) from 1968 to 1982. It was also the conceptual link between the CHM and global distributive justice that made the developed countries, especially the USA, apprehensive about accepting the idea. However, following Pardo's suggestion and with the support of the developing countries, the sea-bed beyond the jurisdiction of states was eventually declared the CHM, thereby establishing the validity of the concept in international environmental law. The concept has since been used to describe other areas outside of the jurisdiction of nation states, such as **Antarctica**, the moon and outer space. During the negotiation of the Climate Change Convention there was a proposal that the global climate should be declared as a CHM. This proposal was, however, rejected by the developed countries. They preferred, instead, that the climate should be seen as the common concern of mankind. The preference of common concern over common heritage is widely seen as an attempt to steer clear from the distributive equity issues which the CHM strongly evokes.

Common Pool Resources (CPR)

Common pool resources (CPR), alternatively called common property resources, are natural or human-made resources over which no single user or decision-making unit has exclusive title. These can include community irrigation systems, rivers, pastures, forests or commonly-owned farmlands. CPR that extend across two or more countries are called transboundary resources, while those that are deemed to be owned commonly by the world community are described as global commons. Examples of transboundary resources include river basins and mountains, while some examples of global commons include the atmosphere, the high seas beyond the jurisdiction of states, the deep ocean sea-bed, **Antarctica** and outer space. CPR are so-called because, by virtue of their nature or size, it is usually difficult to exclude individual users from appropriating them. The technical term for this is to say that they have a low degree of excludability. Another main characteristic of CPR is that they are usually subtractable. This means that the appropriation of such resources by some legal users has the potential to reduce the quality of the resource and limit the ability of others to appropriate it. Of course, the degree of subtractability varies considerably across the various types and sizes of CPR. For some, such as the global atmosphere and the high seas, it was not until recently that humans became aware that these resources have limited carrying capacity. Subtractability is the main feature that differentiates CPR from public goods. Unlike CPR, public goods are said to be non-substractable or non-rivalrous in the sense that the use of these resources does not prevent others from using them or limit their ability to do so. Examples usually included in this category are public roads, street signs, street lights and terrestrial television. However, some argue that this distinction is only theoretical since in the final analysis the use of public goods is still limited to a certain number of people at any one time. Further, they point to the fact that the use of the public roads by some could cause congestion or wear and tear, both of which could limit their use by others. The particular challenge involved in managing CPR is called the **collective action problem**, while the social arrangements that are designed to overcome this problem are called **common property regimes**.

Common Property Regimes

Because **common pool resources (CPR)** usually belong commonly to a group of people rather than to single private individuals, their protection or conservation usually requires some form of more or less explicit agreement and co-operation from the people to whom these resources belong. The social arrangements designed to regulate, preserve or maintain CPR are called common property regimes. In other words, the UN Framework Convention on Climate Change, the European Union's Common Fisheries Policy, the UN Convention on the Law of the Sea and the **Antarctica** Treaty can all be described as global common property regimes. The study of these regimes, how they are developed, how they function and what makes them either effective or non-effective, is a subject of immense interest among scholars of environmental politics. One of the main challenges in the design and operation of common property regimes is how to determine the right amount of resources that can be exploited per time in order to allow users to have optimum benefit from the resource pool without degrading the resource. Other problems include how to allocate the resources equitably to all the potential users, how to make sure that all legal users are effectively represented in the decision-making process and how to resolve conflicts that arise in this process.

Convention to Combat Desertification (CCD)

The Convention to Combat **Desertification** (CCD) was adopted on 17 June 1994 in Paris, France, after a long period of agitation by developing countries in the arid, semi-arid and dry humid zones, particularly in Africa. The convention has the objective of raising awareness of the problem of desertification in the developing countries of the world and of providing a forum for a global and integrated approach to solving the problem. The ultimate aim of parties is to combat desertification and mitigate the effects of drought in countries experiencing such problems. The CCD was negotiated in keeping with a UN General Assembly Resolution which drew from the recommendation made in **Agenda 21** during the **Earth Summit** held in Rio de Janeiro, Brazil, in 1992. Agenda 21 amplified the voice of the affected developing countries by declaring that the problem of drought and desertification requires urgent international attention in the context of global **sustainable development**. Prior to this time, many in the developed countries had resisted the idea of having a UN convention to combat desertification. They argued that desertification is a localized problem which is best addressed by national actions or, at most, through bilateral treaties. They also expressed their concern to avoid a convention whose aims and objectives overlap or conflict with existing conventions, such as those on climate change and **biodiversity**. The developing countries, on the other hand, argued that the problem of desertification is global as it affects all regions of the world. Moreover, they pointed out that the problem of desertification is firmly linked and ultimately exacerbated by the working of the broader international economic and socio-political structure under which the vulnerable countries are mostly disadvantaged. The convention attempts to strike a compromise between the two different positions. It recognizes that the problem of desertification is global and linked to other environmental problems, but also that the affected countries will have to take the lead in dealing with the problem. It also recognized that these countries will require the help and support of the international community in the areas of technological assistance, capacity-building and financial resources. The convention now has over 179 members as parties. However, its political significance and effectiveness remain doubtful.

D

Debt-For-Nature-Swaps

This is a series of agreements between indebted countries and their creditors (governments or **non-governmental organizations—NGOs**) which usually involve the cancellation of debts in return for nature conservation programmes in the indebted countries. This initiative was started by some environmental NGOs in the West in the mid-1980s. The process usually involves the NGOs buying out some portion of the debt of a developing country at a discounted rate. The indebted country then makes available an equivalent sum of money in local currency which is devoted to the management of designated conservation areas. The first of these agreements was negotiated in 1987 between Conservation International (CI), a US-based NGO, and the government of Bolivia. CI bought out a debt of US $650,000 owed by the Bolivian Government for $100,000. The Government of Bolivia in turn promised to make available the equivalent of $250,000 in local currency for the conservation of the Beni Biosphere Reserve which is a place of immense **biodiversity**. In 1988 the World Wide Fund For Nature (WWF) negotiated a similar deal with the Government of the Philippines. The organization bought out a debt of $390,000 owed by the Philippine Government for $200,000 (51% of the face value). In return, the Government agreed to redeem the debt from the NGO by dedicating the equivalent of $390,000 in Philippine pesos for the conservation of the National Park. Over the last two decades other governments, including those of Peru, Ecuador, the Democratic Republic of the Congo and Costa Rica, have negotiated similar deals. It is suggested that, in total, over $180m. has been acquired at a cost of $47m., and that these deals have generated about $130m. for conservation in different areas in the developing countries. The problem with these programmes is that governments sometimes delay or fail to honour their commitments under the deals. At the same time, there are some who argue that these initiatives mainly aim to secure the conservation interests of some in the developed countries but do not engage with the pressing problems of poverty and underdevelopment in the developing countries.

Deep Ecology

A phrase coined by Arne Næss, a Norwegian philosopher and mountaineer, in 1973 to distinguish a variety of environmental concern or movement that entails 'deep' appreciation of humans' relationship with non-human nature. In a presentation in Bucharest, Romania, at the Third World Future Research Conference, Næss argued that there are two types of environmental movement. The first one is based on the logic of modern science and contented with minor technological changes and regulations designed to ameliorate **environmental degradation**. Næss called this shallow ecology. The other, he said, was based on a deeper questioning and deeper commitment of humans to conservation based on respect for the inherent value of the non-human world. He called this deep ecology. The distinguishing feature of deep ecology is that it rejects the idea of moral ranking in which it is held that humans have higher moral value than non-human objects. Deep ecology, some-times called biocentric egalitarianism, holds, rather, that all objects, human and non-human alike, are deeply connected and designed to operate as a harmonious whole. Advocates of deep ecology campaign for a departure from mere ecological science where man seeks to dominate nature by the use of his superior intellectual capacity towards ecological wisdom, characterized by respect for nature and the treatment of non-human objects as the moral equals of humans. Deep ecology's roots can be traced back much further, to the thinking and works of people like Aldo Leopold and Henry Thoreau in the 1950s and 1960s. However, it was Næss who specifically coined the term and provided many philosophical arguments to support the position. The idea remains controversial, however, as many people question the philosophical basis for equating the moral worth of humans with objects like rocks, rivers, trees, or with molluscs. Deep ecology has also been criticized by some in the developing countries as an élitist form of environmentalism. The argument is that the function of deep ecology is to secure the preservation of mountains and other exotic places of recreation for the rich in the developed countries without engaging with the types of environmental problems, such as water scarcity, which are more pressing for the majority of people in the developing countries.

Deforestation

The destruction of large areas of forest by the conversion of forest lands to non-forest land uses or through indiscriminate logging of forest trees. Deforestation is one of the major environmental challenges facing the world today. Deforestation is the single most important cause of **biodiversity** loss world-wide. It is also a major contributor to desertification, drought and climate change. In the last 150 years the world has lost over half of the 1,500m. ha of forest that once covered the face of the Earth and most of this loss has occurred in the last three decades or so. It is estimated that about 50,000 ha of forest are destroyed every single week. In the early centuries, most of the forest loss occurred in Europe and North America as a result of industrial development. More recently deforestation has been concen-trated in Asia, Central and South America, and Africa. Many tropical countries, such as the Philippines, Nigeria, Thailand, Indonesia, Malaysia, China and Bangladesh, have lost over 65% of their rainforest in the last 20 years. Almost more than one-half of the world's forest about 500,000 ha lie in the Brazilian Amazon and is also being degraded at an alarming rate. There are several causes of deforestation, including mining, road construction, construction of dams, and oil and gas exploration. However, the two most important causes of deforestation are logging and the conversion of forests to agricultural uses, such as pastures and arable lands. It is thought that up to 4m. ha of forest are destroyed annually as a result of commercial logging. Commercial logging also facilitates the destruction of forests indirectly as formerly impenetrable forests are opened up to multiple assaults from different users following the roads constructed by industrial loggers. Similarly, millions of hectares of forest are lost yearly as a result of 'slash-and-burn' and the conversion of forests into pastures, soya fields and sugar cane plantations. Deforestation also occurs as a result of the acidification of forest land due to **acid rain**. Some suggest that if deforestation continues at the present rate, only about 20% of the world's forest will remain by 2030. This is disturbing because forests perform many important roles in maintaining life's natural balance. Forests are the home to over 50% of all living organisms and over 90% of primates. Forests are the source of over 26% of the world's medicines and there are strong indications that they could provide the source for many more vital ones in the future. The destruction of the forests, therefore, does not only lead to the direct loss of valuable biodiversity, but also to the reduction of the chances that man might have to discover cures for some deadly human diseases. Moreover, deforestation upsets the balance between the natural carbon and nitrogen cycles which are indispensable to the normal functioning of life. Deforestation also leads to the drying up of aquifers and to a reduction of ground waters, as these are normally recharged by forests. Also, when forest cover is removed, the soil loses its ability to slow down or trap water running down from nearby mountains. This increases the acceleration of precipitation, which often results in flooding. The problem with controlling deforestation is that most of the world's forests lie within the territories of countries and as such are considered as sovereign or national resources. This makes it extremely difficult to develop an international convention to deal with the problem. Besides, the problem of deforestation is also bound up with the broader problems of poverty and underdevelopment. While the developed countries nudge governments of developing countries towards dealing with the problem, developing countries' governments often complain that they have neither the capacity to monitor vast areas of forest, nor alternative resources to fall back on if deforestation is discontinued. Moreover, they point out that much of the pressure for logging and the demand for logged trees come from the developed countries. The closest the world has come so far to regulating deforestation is by the development of the International Tropical Timbers Agreement (ITTA) and the adoption of the Statement of Sustainable Forest Management at the **Earth Summit** in Rio de Janeiro, Brazil, in 1992. However, many consider that neither of these processes have done much to reduce the rate of global deforestation over the last 30 years.

Desertification

Desertification does not, as some might suppose, mean the natural expansion of existing deserts. Rather, it refers to the degradation of dryland ecosystems, such as arid, semi-arid and dry humid areas. The main cause of desertification is human activities, although these are sometimes facilitated by variation in climatic conditions. Desertification is a serious environmental problem as it affects the lives of over 250m. people and over 65% of the total land area of the planet. Equally, over 1,000m. people in over 100 countries are said to be at risk. Dryland ecosystems are generally vulnerable and require careful management in order to remain productive. However, the majority of these ecosystems are inhabited by the world's poorest populations. For the most part, these populations are unable to use the best irrigation systems available or adopt the effective land management or agricultural practices that are needed to ensure the continuous maintenance of these fragile ecosystems. Indeed, in most cases their patterns of land use as well as their rapid increase put so much added pressure on these naturally vulnerable ecosystems that it leads to their complete degradation. Much desertification results from poor agricultural practices, such as 'slash-and-burn', overuse of pesticides and herbicides, consistent use of parcels of land without allowing them time to regenerate, overgrazing, and population settlement patterns. The main impact of desertification is habitat loss, and with it the loss of valuable **bio-diversity**. At the same time, the adverse effects of desertification on humans are equally very considerable. Often when the lands in the countryside are degraded, people resort to migrating to the cities, but they are hardly able to lead decent lives given that most of them are unskilled and uneducated. Others become refugees as they wander about from place to place looking for alternative settlements for themselves and their families. In the process many die of hunger and starvation. In 1994 a **Convention to Combat Desertification** was negotiated in Paris, France, under the auspices of the UN, but the extent to which this convention can address the problem is not clear since many of the causes are deeply intertwined with the broader issues of underdevelopment and poverty.

E

Earth Summit

This is the informal name for the United Nations Conference on Environment and Development (UNCED) held in June 1992 in Rio de Janeiro, Brazil. This was the biggest and most ambitious world environment and development conference ever. It was held for over 12 days and brought together more heads of government than any meeting in history. There were heads or senior officials of over 179 countries present. The Earth Summit was also attended by a record number of **non-governmental organizations (NGOs)** and over 8,000 journalists. The purpose of the meeting was to adopt a new global strategy for economic development that does not compromise the quality of the global environment. The Earth Summit was in effect an attempt by the world community to design a framework and adopt practical steps to achieve global **sustainable development**. The path to Rio started with the release of the **World Commission on Environment and Development Report** (WCED—the Brundtland report) in 1987. The report contained serious warnings of the growing threat to the Earth's system from **environmental degradation**, pollution, industrial economic development and world poverty, and called on the UN to convene a global conference where these threats and possible solutions might be addressed. Five years later, at the insistence of the UN, the world gathered in Rio with much fanfare and many expressions of optimism designed to encourage world leaders to embrace the path of change required to save the Earth. In the end five main documents were agreed. The first was a 27-point Rio Declaration, which endorsed the commitment to pursue sustainable development and to eradicate global poverty. The second was **Agenda 21**, which is a 40-chapter plan of action on how governments at all levels, businesses and individuals might go about pursuing the objectives of sustainable development. The heads of government also adopted two important conventions: the UN Convention on Climate Change and the Convention on **Biodiversity**. The last document was a statement on sustainable forest management. Although, more than 15 years on, the extent to which the objectives of Rio have been achieved is doubtful, the Earth Summit has remained one of the most significant summits in the history of environmentalism, not least for its role in raising the profile of environmental concern at the global level.

Emissions Trading

Emissions trading is a market-based approach used by governments to control or achieve reductions in the emission of gaseous pollutants. In an emissions trading scheme, a government, in close consultation with companies, scientists and other stakeholders, such as environmental **non-governmental organizations (NGOs)**, sets a limit on the amount of a pollutant that should be emitted within a given time period. After setting the limit (the 'cap'), the government then distributes the total amount of emissions permissible among companies and other emitting entities in the form of allowances or credits, with each allowance representing one ton of the relevant emission. Entities that emit more than their allocated allowances are expected to buy emissions credits from those that pollute less. Conversely, entities that use less than their allocated quota are allowed to sell their credits to those that pollute more. In this way the government is able to control the total emissions of a given pollutant within the system while allowing companies and other polluting entities the flexibility to cut emissions in the ways that best suit them. There are many emissions trading systems around the world. Some of the best known include the US sulphur dioxide trading system established under the Clean Air Act of 1990. The system, which is managed by the US Environmental Protection Agency, sets an overall national limit on SO_2 emissions and requires hundreds of participating facilities to reduce their SO_2 emissions or buy tradable permits from the allocations of those that emit less. The first trades were executed in 1992 at around $300 per ton of SO_2. The system is widely acclaimed for having succeeded in cutting emissions by over 50% between 1992 and 1998. Other examples include the emissions trading systems in volatile organic compounds adopted by the US state of Illinois in 1997, and the 1998 Chile offset scheme for the control of air pollution in Santiago, Chile. More recently, emissions trading in carbon dioxide and other greenhouse gases has quickly gained huge popularity. Trading in these gases followed the adoption of the **Kyoto Protocol** to the UN Framework Convention on Climate Change (UNFCCC) in 1997. Under the protocol, emission-reduction quotas are allocated to the developing countries and nations that emit less than their quota of greenhouse gases are allowed to sell emissions credits to polluting countries. In 2002 the United Kingdom opened the first national scheme in the world for trading greenhouse gases. The scheme, which ran from April 2002 to December 2006, had about 33 companies as direct participants and is reported to have achieved emissions reductions of over 7.2m. tons of CO_2 over its lifetime. However, by far the biggest known emissions trading scheme today is the European Union Emissions Trading Scheme (EU ETS), which was introduced across Europe to tackle emissions of carbon dioxide and other greenhouse gases as part of the effort under the UNFCCC to combat the threat of global climate change. All the 27 EU member states' governments are mandated to take part in the scheme, which commenced on 1 January 2005 and is expected to run in two phases. The first phase will run from 2005 to 2007, while the second phase will run from 2008 to 2012 to coincide with the first Kyoto commitment period. Emissions trading has been lauded in some quarters as one of the best ways to reduce the emissions of pollutants in the most cost-effective way. Proponents argue that the schemes work better than control and command mechanisms in which emissions limit values are imposed on particular facilities in that they allow companies the flexibility to determine where and how best to reduce emissions. However, some critics, mainly environmental justice NGOs, argue that emissions trading, like other market mechanisms, allows big and rich polluters to buy their way out of their pollutions without doing much to reduce emissions. They also argue that emissions trading diverts attention from the wider systemic and collective socio-political changes needed to tackle global environmental problems.

Environmental Degradation

Environmental degradation generally refers to the deterioration of any aspect of the biophysical environment, including air, soil and water quality. It also refers to the contamination or destruction of various biophysical environments, such as forests, grasslands, wetlands, flood plains, coral reefs, etc., as well as the extinction of valuable species of flora and fauna. It is generally considered that environmental degradation has been on the increase globally since the 1960s. The increased rate of the degradation of the human and natural environment is mainly attributed to rapid and uncontrolled industrial processes, as well to poverty and rapid growth in human population. Environmental degradation has been a major concern to a great number of people, organizations and world leaders, especially since the 1970s. This concern informed the convening of the first UN Conference on the Human Environment in Stockholm, Sweden, in 1972. At the conference world leaders proclaimed that the protection of the environment should be a major priority of governments, arguing that such protection is closely linked to the promotion of the well-being of the human population. They also set aside 5 June each year as World Environment Day in order to sensitize the general public to the need to preserve the environment. Since then there have been many other notable efforts by governments and various national and international organizations, as well as individuals, to raise awareness of environmental degradation and the need for greater efforts to preserve the environment. These include the landmark **World Commission on Environment and Development Report**, entitled 'Our Common Future' and published in 1987, and the UN Conference on the Environment and Development (**Earth Summit**) in Rio de Janeiro, Brazil, in 1992. However, despite these efforts and increased awareness of environmental issues, some contend that the rate of environmental degradation across the world has continued to grow over the last three decades.

Environmental Impact Assessment (EIA)

This term describes the process followed to ensure that information about the environmental effects of a developmental project is collected, evaluated and taken into account before the project is allowed to go ahead. It involves a systematic identification and assessment of the possible impacts of the proposed project's physical-chemical, biological, cultural and socio-economic component on the environment. The main purpose of EIAs is to ensure that the likely influence a project may have on the environment is fully identified, understood and taken into consideration by policy-makers in reaching a decision on whether the project should go ahead or not. EIAs were first popularized in the USA from the late 1970s following the adoption of the National Environmental Policy Act (NEPA) in January 1970. The Act signalled increased Federal interest in the promotion of environmental quality in the country and mandated the consideration of the environmental impact of major developmental projects across the states. It was, however, the petroleum companies that first adopted the idea in a bid to cope with tougher legislation over their industrial activities. An important aspect of the concept is a desire to ensure public involvement and participation in arriving at decisions concerning the potential effects of proposed developmental projects on the biophysical and socio-cultural environment. Over the past three decades, EIAs have emerged as an important tool in environmental decision-making in all parts of the world, including both industrial and developing countries. Possible areas of impact include conservation, aesthetics, navigation, recreation, water and air quality, food production, mineral needs, historic properties, food hazards, land use, erosion and accretion, water supply, energy needs, health and safety, property rights and general environmental concerns. In some cases, EIA requirements go beyond the articulation of the potential impact of proposed projects to the consideration of alternative scenarios, including the possibility of not developing the project at all.

F

Friends of the Earth (FoE)

One of the most influential environmental pressure groups in the world, Friends of the Earth (FoE) is also one of the most extensive environmental networks. The group has about 70 national organizations and nearly 1.5m. members and supporters world-wide. The group specializes in lobbying governments and politicians, as well as working to increase public awareness of a series of environmental issues. FoE has been very active in campaigning against the construction of new nuclear power stations in Europe and in establishing safety levels for existing nuclear facilities. The organization has also been influential in its campaigns in the areas of air, sea and land pollution, **acid rain**, and climate change. Other notable areas include whaling and the use of whale products, use of ozone-depleting substances, leaded petrol, food additives and non-returnable bottles. The group is noted for its in-depth research and ability to mobilize local people in attempts to influence changes in governments' policies. FoE was originally formed in 1971 by four organizations in the United Kingdom, France, Sweden and the USA. Currently, however, the organization is a federation of about 68 environmental groups, with over 5,000 local activist networks. The organization is highly decentralized, with most of the member groups retaining huge autonomy over their campaign focuses, strategy and budget. For the most part these groups are already well established before joining the federation. Most of these groups call themselves 'Friends of the Earth' in their own languages, while others go by different names. A biannual general meeting elects an executive committee, which oversees the co-ordination of the activities of member organizations. FoE's international secretariat is located in Amsterdam, the Netherlands.

Friends of the Earth International
Prins Hendrikkade 48, POB 19199,
1000 GD Amsterdam, Netherlands
Tel: (20) 6221369
Fax: (20) 6392181
Internet: www.foei.org

G

Genetically Modified Organisms (GMOs) and Crops

One of the most significant advances in modern biological sciences over the last few years has been in the increased ability to manipulate genes and chromosomes in living organisms to create different varieties of life. This process is sometimes referred to as 'genetic engineering'. Recent breakthroughs in the science of genetic modification now enable scientists to create novel forms of life by rearranging genes in living organisms or by transferring some chromosomes from one species to another. Scientists are also now able to transfer genes from plants to animals and vice versa. The most common form of genetic modification process involves the extraction of DNA from one organism, the modification of this DNA in the laboratory, and the subsequent insertion of the modified DNA into another organism's genome in other to produce new useful traits and phenotype order. Crops that are produced through this process of gene alteration are called genetically modified (GM) crops. The first set of GM foods became available in the shops in Western countries in 1990, initially in the USA, and shortly after in Western Europe and Australia. The most popular of these were corn, cotton oil and soybeans, of which up to 60 different crops have now been developed. The first food to be made from a GM ingredient—a vegetarian cheese—went on sale in the United Kingdom in 1992. Since then some Latin American countries, for example Brazil, Argentina and Chile, have begun to plant GM crops on a large scale. In 2001 GM plants reportedly accounted for 46% of the global soybean crop, and 20% of the cotton crop. It is also reported that between 1996 and 1998 the area planted with GM products increased from 2m. to 28m. ha world-wide. Advances in the genetic modification of creatures have equally been rapid. The first GM insect—the pink bollworm—was released in the USA in 2001. This moth was engineered to contain a gene from a jellyfish in the first stage of a genetic experiment designed to eradicate the cotton-destroying pest. But genetic modification is a hugely controversial endeavour. Those who support the genetic modification of plants and animals point to the fact that its processes could lead to the development of crops and animal breeds that have accelerated breeding rates. This is said to be one of the keys to solving the problem of global food shortages in the future. Proponents also point to the advantages in developing crop varieties that are highly resistant to pests and harsh weather conditions. On the other hand, many people caution against the genetic modification process. Those who are more cautious point out that many unintended and unforeseen negative consequences might follow the manipulation and alteration of forms of life in the laboratory. One of the concerns is the possibility that the introduction of GM organisms into natural ecosystems could result in a significant decline in native **biodiversity**. GMOs are common in North America, but not in the European Union (EU). This has created a measure of tension between the world's two major trading blocs. In December 2003 the USA and Canada challenged the EU's *de facto* moratorium on imports of GMOs since 1998 (introduced after widespread public concern about the environmental impact of the plants) in the World Trade Organization. The EU, which now admits some countries already growing GMOs, such as Poland, continues to insist that stringent new rules (finalized in January 2004) are not an infringement of trade. Meanwhile, in February 2004 the USA succeeded in persuading the People's Republic of China to allow imports of GM crops.

Global Climate Coalition (GCC)

Global Climate Coalition (GCC) was the name of the network used by industry and commercial groups in the USA to oppose action against climate change between 1989 and 2002. The group was created in 1989 shortly after the UN inaugurated the **Intergovernmental Panel on Climate Change (IPCC)** to evaluate the evidence linking anthropogenic greenhouse gas emissions to global warming and drastic climate change. Members of the group included Amoco, BP, Dow, Dupont, the American Forest and Paper Association, American Petroleum, Chevron, Chrysler, Exxon, Ford, General Motors, Royal Dutch/Shell, Texaco and the US Chamber of Commerce. Its main purpose was to discredit the science of human-induced climate change as presented by the IPCC, to lobby government (in the USA and beyond) against an international treaty sanctioning mandatory emissions reductions, and to persuade the American public that action against climate change would be harmful to the US economy. The group operated mainly by paying for large public relations campaigns that portrayed the science of climate change as being full of uncertainty. It also commissioned many economic studies that usually concluded that it would be prohibitively expensive to reduce greenhouse gas emissions. The group sponsored extensive advertisement in the US media, claming that Americans would pay 50 cents more for every gallon of gasoline if the USA supported international action against climate change under the **Kyoto Protocol**. In the run up to the 1992 **Earth Summit** the group claimed that increased levels of carbon dioxide in the atmosphere would lead to increases in food production world-wide which would help feed the world's growing poor population. It is widely acknowledged that the group played a significant role in discouraging the USA from supporting the Kyoto Protocol in 1997, and subsequently in influencing the stance of the Bush Administration on the issue of climate change. However, by the late 1990s some members of the group, led by BP/Amoco, had begun to withdraw from the network. BP noted that its withdrawal was based on the conviction that political action against climate change need not wait until the time when the link between anthropogenic greenhouse gases and climate change has been conclusively proven; and of the need to consider policy dimensions, given the possibility that a link might exist. After the withdrawal of BP, other companies, such as Royal Dutch/Shell, Texaco, Dupont, Ford and General Motors, likewise withdrew from the coalition and began to offer qualified support for international action against climate change. This involves mainly stressing the need for voluntary rather than mandatory emissions reductions as well as highlighting the need for a focus on technology- and market-based approaches rather than regulation and other command and control mechanisms. The group finally disbanded in 2002, claiming that it had served its purpose by contributing to perspectives to possible approaches to global warming.

Global Environmental Facility (GEF)

The Global Environmental Facility (GEF) is the largest multinational fund devoted to assisting developing countries to fund projects and programmes that help protect the environment. The GEF is dedicated to six main focal issues, including climate change, **biodiversity**, international waters, land degradation, the ozone layer and persistent organic pollutants. Accordingly, GEF functions as the financial mechanism for the UN Framework Convention on Climate Change, the UN Convention on Biological Diversity, the Stockholm Conference on Persistent Organic Pollutants and the **Convention to Combat Desertification**. GEF is also in charge of disbursing grants to the developing countries under the **Montréal Protocol** for the Protection of the Ozone Layer. GEF was established in 1991 by the UN Environmental Programme, the UN Development Programme and the World Bank, all of which serve as implementing agencies for the Fund. GEF's activities are overseen by an independent board of directors, comprising representatives from 32 constituencies (16 from developing countries, 14 from developed countries, and two from countries with transitional economies). The council meets twice each year for three days to develop, evaluate and adopt GEF programmes.

Greenhouse Effect

Greenhouse effect refers to the natural process by which the Earth is warmed through the retention of some of the sun's energy by the atmosphere. Scientists had long discovered that the naturally occurring functions of some trace gases in the atmosphere mimic those of the glass in a greenhouse, hence the name 'greenhouse effect'. Just as the glass of the greenhouse is transparent to sunlight but opaque to the infrared radiation emitted by the warm surfaces within, so certain atmospheric gases, including water vapour, carbon dioxide, methane and chlorofluorocarbons (known collectively as greenhouse gases), allow sunlight to pass unimpeded but absorb the radiation from the Earth's surface. It is the absorbed radiation that is responsible for the warming of the land, atmosphere and the oceans. The Earth receives energy from the sun mostly in the form of ultraviolet radiation. About 30% of this ultraviolet radiation is reflected directly back to the sun as ultraviolet rays while about 70% is absorbed by the Earth's surface. The absorbed energy is subsequently released gradually by the Earth's surface in the form of infrared radiation, with the intensity of the radiation increasing in proportion to the atmospheric temperature. Hence, the greater the amount of greenhouse gases in the atmosphere, the more infrared radiation is absorbed, leading to an increase in the atmospheric temperature. On average, the atmospheric concentration of greenhouse gases in the atmosphere has increased by more than 15% in the last 100 years or so. Scientists attribute this increase mainly to the burning of fossil fuels such as gas, coal, and oil, which results in the release of large amounts of greenhouse gases into the atmosphere. Scientists believe that the rapid increase in greenhouse gases in the atmosphere is responsible for the increased mean temperature of the atmosphere over the last few decades. This condition is widely referred to as global warming. There are many uncertainties over the rate at which greenhouse gases accumulate and the resulting consequences, but scientists in the **Intergovernmental Panel on Climate Change** have estimated that a doubling of the present level of carbon dioxide concentration (or its equivalent in other gases) could cause a rise in the average temperature of $1°C–3.5°C$, and that such a doubling may occur by the mid-to late 21st century unless steps are taken to reduce the emission of greenhouse gases. Scientists say that in order to avert dangerous changes in the climate, it is necessary to cut global emissions of greenhouse gases by up to 50% of the 1990 level. However, under the **Kyoto Protocol** agreed at a conference of parties to the UN Framework Convention on Climate Change in December 1997, nations agreed to reduce emissions of greenhouse gases at the global level by about 6.2% by 2012. Most approaches to emissions reduction involve increasing the efficiency with which energy is used, as well as the use of **renewable energy** sources as alternatives to fossil fuels.

Greenpeace International

Greenpeace International is one of the most influential environmental **non-governmental organizations** in the world. Greenpeace has consistently played an active role in campaigning and influencing government policies on a number of environmental issues, including recycling, the elimination of nuclear power and nuclear weapons, marine protection, inappropriate disposal of toxic waste, etc. Greenpeace is particularly noted for its passion and unique campaign methods, which involve strategic direct but non-violent confrontation of those engaging in the practices it campaigns against. The organization was founded in 1971 and has over 2.1m. members and supporters. The group is present in about 40 countries across Europe, the Americas and Asia and the Pacific, with its headquarters in the Netherlands. As a policy Greenpeace does not receive money from governments but relies solely on donations from supporters and grants from foundations.

Greenpeace International
Ottho Heldringstraat 5,
1066 AZ Amsterdam, Netherlands
Tel: (20) 7182000
Fax: (20) 5148151
Email: supporter.services@int.greenpeace.org
Internet: www.greenpeace.org

G77

A short form for the Group of 77, which is the name for the umbrella group of developing countries which serves as a platform for the articulation of views on issues of common concern, especially with respect to deliberations within the UN or UN-sponsored treaties. In addition to articulating views, the G77, being an officially recognized intergovernmental organization, frequently sponsors motions during UN deliberations. It also produces joint declarations, action programmes and agreements on issues pertaining to global development, political economy and environmental sustainability. G77 is generally regarded as the official mouthpiece of the developing countries in global environmental negotiations. The forum was very active during the third UN Conference on the Law of the Sea (UNCLOS III, 1968–82), regularly providing a basis for internal consultation and the strengthening of developing countries' position on a number of thorny issues. The forum has also been very active in the UN conferences on climate change, including the Framework Convention. It has provided the platform on which the less industrialized countries have been able to highlight issues of justice and equity implicated in climate change. The G77 was formed in June 1964 by 77 developing countries that are signatories to a joint declaration issued at the end of the first session of the UN Conference on Trade and Development (UNCTAD) held in Geneva, Switzerland. There are now about 130 developing countries that are members of the group.

H

Helsinki Convention

The more popular name for the Convention on the Protection of the Marine Environment of the Baltic Sea Area. The convention is named after the Finnish capital, where the treaty was first signed in 1974 by the then seven Baltic states. The convention works to regulate all sources of pollution of the Baltic Sea as well as to restore and safeguard its ecological balance. The convention covers the whole of the Baltic Sea area, including inland waters as well as the water of the Sea itself and the sea-bed. Measures are also taken in the whole catchment area of the Baltic Sea to reduce land-based pollution. The 1974 convention came into force on 3 May 1980. In 1992 the convention was extensively revised in the light of political changes and developments in international environmental and marine law. The new convention, which has a wider remit, came into force on 17 January 2000. The 1992 convention was signed by nine Baltic states (Denmark, Estonia, Finland, Germany, Latvia, Lithuania, Poland, Russia and Sweden) as well as the European Union. The convention has a governing body—the Helsinki Commission (HELCOM)—which develops and recommends policies to the contracting parties, supervises implementation of rules and acts as a focal point for information for the member states.

I

Intergovernmental Panel on Climate Change (IPCC)

The Intergovernmental Panel on Climate Change (IPCC) is a body established by two UN agencies: the World Meteorological Organization (WMO) and the **United Nations Environmental Programme (UNEP)** in 1988 to assess scientific, technical and socio- economic information relevant for the understanding of climate change, its potential impacts and options for **adaptation** and mitigation. The IPCC does not itself carry out scientific research on climate change, nor does it engage in the monitoring or gathering of raw climate data. Rather, the panel bases its assessments on peer-reviewed and published scientific literature and technical reports. The bulk of the members of the IPCC comprise government scientists, but a significant number of independent academic scientists, researchers and scholars from various countries of the world also participate regularly in its activities. Over 2,500 scientific experts and 850 authors contribute to the works of the IPCC. The panel is open to all members of the UN and the WMO. The first meeting of the IPCC took place in November 1988 in Geneva, Switzerland, where three expert or working groups were set up. Working Group I has the function of assessing the scientific aspects of the climate system and climate change. Working Group II assesses the impact of climate change on socioeconomic and natural systems while Working Group III assesses policy options for limiting greenhouse gas emissions as well as options for adaptation. Each working group has two co-chairs, one from the developed and the other from the developing world, as well as a technical support unit. The first reports (called Assessment Reports) of the three working groups were presented at an IPCC meeting in Sweden in August 1990. These reports played a significant role in galvanizing activities and government support for the negotiation of the UN Framework Convention on Climate Change (UNFCCC) between 1990 and 1992. The IPCC reports are updated every five years. The second Assessment Reports were published in 1995; the third in 2001 and the fourth in 2007. To date, the IPCC reports, despite some controversies relating to objectivity and bias, continue to provide the basis for many governments' position in negotiations on climate change. For example, the Second Assessment Reports provided key input to the negotiations that led to the adoption of the **Kyoto Protocol** to the UNFCCC in 1997. Apart from these comprehensive reviews, the IPCC also produces hundreds of special reports, technical papers, summary reports and CD ROMs on various aspects of climate change. The governing body of the IPCC is a panel that meets in plenary sessions about once a year. The panel adopts IPCC reports and decides on work plans for the working groups, the structure of reports and on budgetary matters and procedures. A secretariat, which is hosted by the WMO in Geneva, oversees the day-to-day operations of the group.

International Human Dimensions Programme on Global Environmental Change (IHDP)

The International Human Dimensions Programme on Global Environmental Change (IHDP) is a global network of social scientists concerned with the promotion of research on the human dimensions of global environmental change. The areas of coverage include land use, urbanization, human security, interactions between institutions and the global environment, sustainable production and consumption patterns, food and water issues and climate change. The IHDP was founded in 1996 by the International Council for Science (ICS) and the International Social Science Council (ISSC) as part of a collaborative effort to promote sound research and the application of key findings of research to help address prevalent global environmental challenges. The IHDP functions to bring scientists working in various aspects of the environment together and to provide the platform for interactions and collaboration among such scientists. The IHDP equally works to formulate new research agendas, stimulate high-quality research and to synthesize existing bodies of knowledge in the various aspects of the human dimensions of the environment. As part of its efforts to achieve these objectives the IHDP sponsors conferences, training and capacity-building work-shops, as well as the publication of quality research findings. At present there are about seven core projects under the IHDP. These include Global Environmental Change and Human Security (GECHS), Institutional Dimensions of Global Environmental Change (IDGEC), Industrial Transformation (IT), Land Use and Land Cover Change (LUCC), Urbanization and Global Environmental Change (UGEC) and Global Land Project (GLP). The activities of IHDP are guided by an international science committee made up of reputable social scientists from various disciplinary backgrounds and countries of the world. There are also about 32 national committees, which determine priorities and co-ordinate both country and region-specific research in the areas of human dimensions of global environmental change. The international secretariat of the IHDP is hosted by the Bonn Campus of the UN University, which has been a joint sponsor of the programme (with ICS and ISSC) since March 2007.

International Maritime Organization (IMO)

The International Maritime Organization (IMO) is a specialized agency of the UN concerned with the international regulation of shipping and the preservation of the marine environment. There are about 26 conventions operating under the

IMO, with at least seven in the area of marine pollution. However, the best-known of these are the International Convention for the Prevention of Pollution from Ships (MARPOL) and the Convention on the Prevention of Marine Pollution by Dumping of Wastes and Other Matter (LDC). MARPOL was adopted on 2 November 1973. The convention covers pollution by wide-ranging substances, including oil, chemicals, harmful substances in packaged form, sewage and garbage. In 1978 a protocol was incorporated into the convention to address the issue of accidents involving tankers. Despite the importance of the convention in the prevention of marine pollution, the rate of ratification of MARPOL was originally very slow as many shipping nations struggled to meet the new requirements and specifications for waste discharge stipulated by the convention. In a bid to reduce the incidence of oil pollution from ships, the convention recommended a series of oil discharge criteria for different kinds of ships and required that certificates of clearance be obtained as proof that engine specification and operating standards had been met. Furthermore, the convention introduced the concept of 'special areas' (areas of the sea that are considered to be especially vulnerable) where oil discharges are completely prohibited. The 1973 convention identified the Mediterranean Sea, the Black Sea, the Baltic Sea, the Red Sea and the Persian (Arabian) Gulf as special areas. All oil-carrying ships are required to be capable of operating the method of retaining oily wastes on board through the 'load on top' system or for discharge to shore reception facilities while plying the special areas. The convention also has detailed standards on packing, labelling and methods of carriage that are designed to prevent pollution by other harmful substances from ships. Any violation of the MARPOL 73/78 Convention within the jurisdiction of any party to the convention is punishable either under the law of that party or under the law of the flag state. The Convention on the Prevention of Marine Pollution by Dumping of Wastes and Other Matter (otherwise known as the London Convention) was adopted in London, United Kingdom, in November 1972. The convention has the aim of regulating a common but clearly unsustainable practice of using the sea as a dumping ground for all sorts of household and industrial waste. The convention, which came into force on 30 August 1975, prohibits the dumping of certain hazardous materials in the sea, requires a prior special permit for the dumping of a number of other identified materials and a prior general permit for other wastes or matter. The convention's annexes list wastes that cannot be dumped and others for which a special dumping permit is required. Recent amendments have banned the dumping into the sea of low-level radioactive wastes and introduced the concept of **precautionary principle** which stipulates appropriate preventative measures to be taken when there is reason to believe that wastes or other matter introduced into the marine environment are likely to cause harm, even when there is no conclusive evidence to prove a causal relation between inputs and their effects. It was also agreed under the convention that, from February 2007, the sea-bed could be used for the storage of carbon dioxide. This allowance was given in response to suggestions that carbon dioxide capture and storage would play a significant role in the effort to combat dangerous climate change.

International Maritime Organization
4 Albert Embankment,
London SE1 7SR, United Kingdom
Tel: (20) 7735-7611
Fax: (20) 7587-3210
Email: info@imo.org
Internet: www.imo.org

International Tropical Timber Agreement (ITTA)

The International Tropical Timber Agreement (ITTA) was signed on 28 March 1983 by 69 countries, comprising both producers and consumers of tropical timber. The agreement, which came into force on 1 April 1985, was designed to provide a framework for international co-operation between the producing and the consuming countries with respect to dealings in tropical timber. The agreement has as its main objective the desire to ensure the conservation and development of timber forests in order to guarantee optimum utilization while maintaining the ecological balance of the regions where the timbers are harvested. However, although the need for conservation is mentioned as an objective in a few other articles of the agreement, the bulk of the activity of member states under the agreement has concentrated on the development of the international timber market, including the elaboration of trade rules with the aim of ensuring maximum utilization of tropical timber. Accordingly, many have argued that the ITTA is not a conservation or environmental treaty but purely a trade agreement. This view underpins the continuous agitation by many environmentalists for the development of an international treaty on the conservation and sustainable management of the global forests. The ITTA is administered by the International Tropical Timber Organization (ITTO), which is established by Article III of the ITTA and located in Japan. The ITTO is the corporate entity of the agreement. There is also an International Tropical Timber Council which is the highest decision-making body of the organization.

International Tropical Timber Organization
International Organizations Center, 5th Floor,
Pacifico-Yokohama, 1-1-1, Minato-Mirai, Nishi-ku,
Yokohama 220-0012, Japan
Tel: (45) 223-1110
Fax: (45) 223-1111
Email: itto@itto.or.jp
Internet: www.itto.or.jp

J

Joint Implementation (JI)

Joint Implementation (JI) is one of the so-called flexible mechanisms put in place by the **Kyoto Protocol** to make it easier for industrialized countries to meet their emission-reduction commitments under the protocol. Specifically, JI allows industrialized countries to sponsor or collaborate with Eastern European countries (Countries in Transition—CIT) in the establishment of low-carbon development projects. The idea is that by sponsoring such low-carbon development projects, it will be possible to reduce the amount of emissions that would otherwise have occurred if the CIT were to follow cheaper, high-carbon development pathways. At the same time, it is considered that investing in such projects will provide the opportunity for the transfer of technology while allowing the industrialized countries to make emissions reductions in locations where they would get the highest result per amount of money spent. Accordingly, the policy provides that both the investing and the host country share the emission-reduction credits that accrue from following the more carbon-efficient developmental pathway. The policy is, however, not restricted to deals between developed and CIT countries, but also allows for collaboration between two or more industrialized countries.

K

Kyoto Protocol

The protocol to the UN Framework Convention on Climate Change (UNFCCC) adopted in Kyoto, Japan, on 11 December 1997. The protocol came into force in 2005 following its ratification by Russia. The Kyoto Protocol provides the detailed guidelines and policy mechanisms that allow governments to take more specific and measurable action to combat climate change in line with the objective of the Framework Convention. The most outstanding aspect of the protocol is that it commits industrialized countries to legally, individually binding targets to reduce their greenhouse gas emissions. Under the protocol, the Annex I (industrialized) countries agreed to cut their (cumulative) greenhouse gas emissions by at least 5% from the 1990 base line between 2008 and 2012, which is known as the first commitment period. Remarkably, the protocol allocates to all of the industrialized countries specific emission-reduction quotas based on a number of criteria, including historical emissions, level of technological advancement, extent of dependence on fossil fuels, and peculiar development challenges. For examples, the European Union (EU) has a quota of –8% which it is expected to distribute among its then (as of 1997) 15 member states; the USA has a quota of –7%; Canada, Japan and Hungary each have a quota of –6%; Norway has a quota of +1%; and Australia has a quota of +8%. The Kyoto Protocol also details a series of rules and project-based mechanisms, such as the Clean Development Mechanism (CDM) and **Joint Implementation (JI)**, through which country parties might collaborate to meet their emission-reduction targets ('Kyoto targets'). The protocol also provides for the trading of emissions among Annex I parties as a means of achieving emissions reductions. Despite these provisions for collaboration among parties and for **emissions trading**, the protocol requires contracting parties to put in place strong domestic measures to reduce greenhouse emissions. The protocol provides for the review of its commitments, so that these can be strengthened over time. Negotiations on targets for the second commitment period are due to start in 2005, by which time Annex I parties must have made 'demonstrable progress' in meeting their commitments under the protocol. Only parties to the convention can become parties to the protocol, but this involves a separate ratification process. At present about 175 parties have ratified the protocol. The USA is opposed to the protocol and has indicated its intention not to ratify.

L

Law of the Sea Conferences

A total of three UN conferences designed to elaborate international laws regarding different aspects and uses of the world oceans and sea. The first of these conferences—the first UN Conference on the Law of the Sea (UNCLOS I)—took place in Geneva, Switzerland, in 1958. The second (UNCLOS II) was held in the USA in 1960; and the third (UNCLOS III), which took over eight years to negotiate, was signed in Montego Bay, Jamaica, in 1982. Some of the areas covered include the limits of the territorial sea and the continental shelf, the passage of ships through the territorial sea, the nature of rights and conditions for exploration on the continental shelf, the limit of the Exclusive Economic Zone (EEZ), the rights of landlocked and geographically disadvantaged states, prevention of pollution of the high seas, ocean and sea-bed mining and conservation of the living resources of the sea. All through the 16th and 17th centuries, the prevalent thought governing the use of the world's seas and oceans was that these resources belonged to no one in particular and were thus subject to use by everyone. This thought was generally captured in the doctrines of mare liberum (the freedom of the sea) and *res nullius* (the sea belongs to no one), both articulated and advocated by leading thinkers of the time, especially Dutch lawyer Hugo Grotius. Apart from a narrow strip of water generally regarded as the natural extension of the national territory of coastal countries (hence the name territorial sea), the vast portion of the sea was deemed as belonging to no one in particular and subject to use by anyone on the basis of first come, first served. Following, however, a number of developments in the late 17th and early 18th centuries, such as advances in technology, need for pollution control, population growth and security concerns, many coastal countries began to express the need for the extension of their territorial seas. This condition soon led to claims and counter-claims over various portions of the sea by neighbouring coastal states, as well as threats of seizure and confiscation of military and commercial ships. Desirous to forestall a breakdown of law and order, the UN convened the first Conference on the Law of the Sea in 1958 to determine the limit of territorial sea and to elaborate laws regarding other aspects of the ocean. In UNCLOS I four conventions were agreed. These included a convention on the territorial sea and the contiguous zone, a convention on the continental shelf, a convention on the high seas and a convention on fishing and conservation of the living resources of the sea. Although UNCLOS I is generally regarded as a success, the issue of the limits of the territorial sea could not be determined. In 1960 UNCLOS II was convened to address this issue, but the conference failed to resolve the matter. Following a moving speech by the Permanent Representative of Malta at the UN in December 1967 regarding the abundance of precious minerals in the sea-bed and the need for conservation and pollution prevention, the General Assembly agreed to organize a third conference on the law of the sea to elaborate a more comprehensive law on the use of the sea, as well as to resolve the outstanding issue of the limits of the territorial sea. The negotiation of UNCLOS III lasted from 1968 until 1982, when it was concluded in Montego Bay, Jamaica. UNCLOS III has been described as the most comprehensive conference ever organized by the UN. The convention has more than 320 articles and nine annexes (some of which themselves have up to 40 articles). The conference entered into force on 16 November 1994 on the deposition of the 60th instrument of ratification by Guyana. As of 18 October 2006, a total of 150 states and entities (such as the European Union) are parties to the convention. The USA is the only member of the UN Security Council and NATO that has yet to ratify the convention. The USA is opposed to the conference because it declares that the sea-bed beyond the jurisdiction of states is a **common heritage of mankind** and that the mining of seabed minerals should be supervised directly by the UN rather than subject to appropriation by the industrialized countries.

Limits to Growth

A study commissioned by the **Club of Rome** and published in 1972 by a team of economists working with Prof. Dennis Meadow of the Massachusetts Institute of Technology in the USA. The report predicted that the world as we know it would come to an end in about 100 years if the patterns of population growth and consumption of non-renewable resources remained unchanged. The team came to this conclusion by examining the behaviour mode of five major factors that affect quality of life using a computer model nicknamed World III. The factors investigated include industrialization, population growth, pollution, food production and depletion of non-renewable resources. The report noted that there was exponential growth in the amount of all the basic elements considered in the model. While it acknowledged that the computer model was not perfect, the report none the less claimed that there were sufficient grounds to conclude that there would be a sudden and uncontrollable decline in both population and industrial capacity within a century as the world system reaches the limits to growth. The report predicts that the limits to growth will occur as a result of the combined effect of the initial growth in population, depletion of non-renewable resources and an increase in rates of pollution, all of which will lead to decreases in productivity, even within the context of improved technology. The report generated serious controversy around the world. Within a short space of time it was translated into over 15 languages and sold over 30m. copies. It also played a major role in activating wider and serious debates on global **sustainable development**. The report itself proposed a series of changes and alternative development pathways that could be followed in order to avoid the disaster as predicted. It proposed zero population growth, a levelling off on industrial production, pollution control, recycling of materials, the manufacture of more durable and repairable goods and shift from consumer goods to a more service-oriented economy.

Long Range Transboundary Air Pollution Convention (LTRAP)

An environmental treaty signed in Geneva, Switzerland, on 13 November 1979 to deal with the problem of transboundary air pollution among countries in Europe, Central Asia and North America. The convention entered in force on 16 March 1983. The UN Economic Commission for Europe (ECE) convened negotiations to develop the 1979 convention in response to the problems with transboundary pollution and **acid rain** deposition in Europe. Negotiations were facilitated by the International Institute for Applied Systems Analysis (IIASA), which created a model demonstrating patterns of long-range transport of air pollutants and the acid deposition arising in the process. LTRAP has been considered as one of the most effective modern-day environmental treaties as it has led to a drastic reduction in emissions of most of the air pollutants it aims to control. Initially targeted at reducing the effects of acid rain through control of emissions of sulphur, its scope has since been widened to include nitrogen pollutants, volatile organic compounds and photochemical oxidants. Heavy metals and persistent organic pollutants were subsequently also added. The protocols signed to date include the first **Sulphur Protocol** (1985; entered into force 2 September 1987); the protocol on the control of nitrogen oxides or their transboundary fluxes (NOx protocol, 1988; entered into force 14 February 1991); the protocol concerning the control of emissions of volatile organic compounds (VOC) or their transboundary fluxes (the VOC protocol, 1991; entered into force 29 September 1997); the second **Sulphur Protocol** (1994); protocols to reduce emissions of heavy metals (1998; entered into force 29 December 2003); and the protocol on persistent organic pollutants (1998; entered into force 23 October 2003). The most recent agreement under the LRTAP Convention is the Protocol to Abate Acidification, Eutrophication and Ground-level Ozone—also called the multi-effect or Gothenburg protocol. It was formally adopted in Gothenburg, Sweden, in 1999 and has been signed by 31 countries.

Love Canal

The name of a housing estate in Niagara City, New York State, USA, which eventually became one of America's worst environmental tragedies in history. Love Canal is also one of the key events that provided the impetus for the formation and spread of grassroots environmental justice movements in the USA. In the early 1910s a man named William T. Love had a vision to generate cheap electricity by constructing a canal between the higher and lower Niagara Rivers. The project, like the canal, was never completed, as cheaper methods of transporting alternative electric currents for long distances were discovered. In the 1920s the canal was turned into a municipal and industrial chemical waste dump. In 1953 Hooker Chemical Co, the then owner of the property, covered up the canal with earth and sold it to the city for about US $1. The city went ahead in the late 1950s to convert the site into a housing estate. However, unknown to everyone, the chemicals buried under the earth were gradually seeping to the surface and contaminating the air, food, and water sources in the estate. By 1978, however, the problem had become a massive tragedy, with several instances of deaths, miscarriages, birth defects, skin burns and other sorts of illness developing among residents. Following a particular incidence of unusually long rainfall, corroding waste-disposal drums were seen breaking up through the grounds of backyards, trees and gardens were turning black and dying, puddles of noxious substances were seen in many areas in the estate, including in people's basements and school grounds. When the scale of the problem became evident, all of the residents were evacuated and the homes were repurchased by the city. The US President and the US Senate also approved the release of Federal aid to help resettle the victims. Later, the Federal Government filed a suit against Hooker Chemical Co and its parent corporation, Occidental Petroleum Corpn, requesting that the company be mandated to clean up the chemical waste dump. The suit sought a total of about $117m. in clean-up costs from Hooker as well as reimbursement for more than $7m. spent by Federal agencies in emergency measures at Hooker's Love Canal waste disposal site. After a protracted legal battle the parties settled out of court, with the chemical company agreeing to commit about $20m. to clean-up costs. The site has since been cleaned up and declared free from contamination, but the event has continued to provide some sort of reference to environmental justice campaign groups in the USA up to the present.

M

Montréal Protocol

The Montréal Protocol was a protocol negotiated by the contracting parties to the **Vienna Convention for the Protection of the Ozone Layer** in a bid to take collective and practical steps to combat the problem of ozone-layer destruction. The protocol was signed in Montréal, Canada, on 16 September 1987 and entered into force on 1 January 1989. The protocol was signed by 118 parties, including the European Community, now the European Union, and by the end of 2002 every single party that signed the protocol had ratified it. The protocol is often regarded as a significant example of an effective multilateral environmental treaty. Parties to the protocol agreed first to control and subsequently to gradually phase out the production and consumption of a number of chlorofluorocarbon gases implicated in the destruction of the stratospheric ozone layer. Sections of the protocol made special provisions for the developing country parties to the convention. Specifically, countries whose annual calculated level of consumption of the controlled substances was less than 0.3 kg per caput on the date of the entry into force of the protocol were allowed to delay their compliance with the control measures by ten years in order to enable them to meet the basic domestic needs of their populations. Parties also agreed on a number of technical assistance measures to enable developing country parties to phase out the production and use of the chemicals in question in the shortest possible time. On 29 June 1994 an important amendment was made to the protocol (the London Amendment). The amendment, which entered into force on 10 August 1992, provided for the inclusion of 12 new chemicals on the list of controlled substances and 34 new chemicals on the list of substances with reporting requirements. The London Amendment also established a financial mechanism which included the establishment of a multilateral fund to assist developing countries to comply with the control measures.

Multilateral Environmental Agreements (MEAs)

Multilateral Environmental Agreements (MEAs) are internationally negotiated agreements designed to protect various aspects of the environment. Some of the earliest MEAs include the Migratory Bird Treaty between the USA and Great

Britain (for Canada), signed in 1916; the Western Hemisphere Convention on Nature and Wildlife Preservation—a 1940 treaty that provides a policy framework for the USA and 17 other American republics to 'protect and preserve in their natural habitat representatives of all species and genera of their native flora and fauna, including migratory birds'; and the 1959 Antarctic Treaty, which was designed to foster scientific co-operation among consultative members and to adopt measures for the protection of the native birds, mammals, and plants of the Antarctic. Others include the 1911 Convention Respecting Measures For The Preservation And Protection Of The Fur Seals In The North Pacific Ocean; and the 1958 UN Conference on the Law of the Sea. However, since 1972 there has been dramatic growth in the number of multilateral environmental agreements. This has been attributed to a number of factors, such as the increased rate of global **environmental degradation**, increased awareness of these problems and the increasing interdependence among the countries of the world. According to some sources, there are now well over 749 multilateral environmental agreements, including about 405 treaties, 152 protocols and about 236 amendments. As the number of environmental agreements has increased, so have their scope and their complexity. Whereas some of the earliest known environmental agreements were concerned with the conservation of non-human nature, more recent multilateral environmental agreements now involve regulations of global trade. Examples include the Convention on International Trade in Endangered Species of Wild Flora and Fauna (**CITES**); the Convention for the Control of Transboundary Movements of Hazardous Wastes and their Disposal (the **Basel Convention**). Others, such as the UN Conference on the Law of the Sea and the UN Framework Convention on Climate Change, entail far-reaching changes in the socio-economic structure of nations. Yet, there are some who argue that the rate of environmental degradation and the overall state of the global environment continue to be a major concern, despite the ever-growing list of multilateral environmental agreements in the last three decades.

N

Non-governmental Organizations (NGOs)

Non-governmental organizations (NGOs) are private, voluntary, non-profit-making organizations that act in the public interest on wide-ranging issues, such as human rights, child welfare, poverty alleviation, humanitarian aids, women's rights, etc. NGOs educate the public on their issues of focus. They also sometimes lobby governments to act or make favourable policies with regard to the issues of concern. NGOs also mobilize the public to action and often function as advocates on behalf of people on given issues. Environmental NGOs have played a very crucial part in the development of environmental awareness and in the development of key environmental policies, as well as **multilateral environmental agreements (MEAs)** across the globe. **Greenpeace** played a crucial role in the development of the **Basel Convention** on the Transboundary Movement of Hazardous Wastes and their Disposal, **Friends of the Earth** played an important role in getting bottling companies in Europe to package their products in recyclable bottles, and conservation organizations, such as the **World Conservation Union**, have been active in promoting various MEAs since the 1960s. Throughout the world, there are more than 7,000 international NGOs that are concerned with the environment and development, with millions of supporters. In all it is estimated that globally there are well over 20,000 NGOs ranging from small grassroots agencies to influential international groups like Greenpeace and Friends of the Earth. The growth of environmental NGOs has been very dramatic ever since the UN Conference on the Human Environment in Stockholm, Sweden, in 1972. This growth is often said to be a result of the increased rate of **environmental degradation** world-wide and the increase in socio-economic globalization, which has increased interaction among different groups across the world, as well as changes in the functions of government, which allow private organizations to play more active roles in governance.

O

Oslo Convention (OSPAR)

The Oslo Convention (OSPAR) is a short name for the Convention for the Protection of the Marine Environment of the North-East Atlantic (also known as the OSPAR Convention). This is the treaty originally designed to provide the basis for the protection of the North Sea from pollution arising from the dumping of certain wastes from ships and aircraft, and wastes resulting from oil-drilling activities. The convention was signed in Oslo, Norway, on 15 February 1972 and entered into force in 1974. The signatories are mainly countries bordering the North-East Atlantic, such as Belgium, Denmark, Finland, France, Germany, Iceland, Ireland, the Netherlands, Norway, Portugal, Spain, Sweden and the United Kingdom. Annex I of the treaty provides a list of wastes whose dumping is strictly prohibited by the convention (the 'black list'). These include all organo-halogen compounds; organosilicon compounds; mercury and mercury compounds; cadmium and cadmium compounds; as well as persistent plastics and other persistent synthetic materials. Annex II contains the list of chemicals requiring special care and for which special permission must be sought before dumping them in the sea (the 'grey list'). These include arsenic, lead, copper, zinc and their compounds, cyanides, fluorides, and pesticides and their by-products. Overall, the Oslo treaty has similar black lists and grey lists to the London Convention of substances that are banned from dumping. In 1992 this convention was combined with the 1974 Paris Convention on land-based sources of marine pollution to form a comprehensive OSPAR Convention, which entered into force on 25 March 1998. More recently the convention has also imposed certain restrictions on the use of marine incineration and dumping of sewage sludge, as well as the use and discharge of polychlorinated biphenyls. Supervision of the implementation of OSPAR and monitoring activities are carried out by the OSPAR Commission, which consists of the representatives of each contracting party.

Ozone-layer Depletion

Ozone is the name of a molecule that consists of three atoms of oxygen stuck together, in contrast to the oxygen in the atmosphere, which has two atoms of oxygen. It occurs naturally in small amounts in the Earth's upper atmosphere, and in the air of the lower atmosphere after a lightning storm. In the lower atmosphere (the troposphere) the ozone molecule is highly unstable and reactive. Even at low concentrations it is irritating and toxic and causes many health problems when inhaled. However, in the higher atmosphere (the stratosphere) the gas is far more stable and binds together to form a protective layer that blocks the harmful ultraviolet rays of the sun from reaching the Earth. This layer is found between 15 km and 30 km above the Earth. In 1984 scientists working from the British Antarctic Survey (BAS) base at Halley Bay observed that a large portion of the ozone layer had deteriorated to leave a massive hole in the layer. Scientists have discovered that the destruction of the ozone layer is caused by man-made gases, particularly chlorofluorocarbons (CFCs) and halons. CFCs have been used in a wide variety of manufacturing steps and products, including as refrigerants, solvents in the electronics industry, foaming or blowing agents, aerosol propellants, fire extinguisher agents, dry cleaning solvents, degreasing agents, key components in making rigid foam insulation for houses and household appliances, and foam packaging insulation material. These gases had been favoured because they are highly inert and therefore pose no danger to human health. However, over time they disintegrate under the rays of the sun and release chlorine radicals in the process. The chlorine radicals released are able to trigger a series of chain reactions with the ozone molecules which causes their destruction, without being themselves consumed. For example, one chlorine atom can cause about 100,000 destructive reactions before eventually leaving the atmosphere. Ozone-layer destruction is particularly significant in the **Antarctica** region because a combination of a number of complex meteorological factors, including deep temperatures in the polar night, ice cloud formation, followed by the polar sunrise. It is for this reason that the ozone hole is most noticeable over Antarctica between September and October each year. The Antarctic hole now measures about 23m. sq. km, nearly the size of North America. The use of CFCs has been banned under the **Montréal Protocol** on Substances That Deplete the Ozone Layer. However, scientists still predict that it will be several decades before the hole in the ozone layer could be expected to recover. There are, however, doubts whether this could ever happen as climate change is thought to have an adverse effect on the recovery process.

P

Polluter Pays Principle (PPP)

The Polluter Pays Principle (PPP) is an international environmental law designed to ensure that the full cost of the damage caused by pollution is borne by the party that is responsible for causing the damage, that is, the polluting party. The principle was originally advocated by the Swedish Government and has since received strong support from the Organization for Economic Co-operation and Development (OECD). It is fair to say that the principle is widely regarded as a cardinal norm in environmental decision-making in Europe and North America, although its application in global environmental governance circles has been very patchy. The policy was put in place in response to concerns that most polluters, especially companies and industries, regularly pass on the cost of their pollution to the government or society at large. A basic tenet of the principle is the idea that companies should internalize the environmental cost of their production processes. What this means is that companies should ensure that the prices of goods and services reflect the full cost of production, including the environmental (social) costs. The principle has not been easy to enforce, particularly where atmospheric pollutants are concerned, because there are so many sources and it is not always possible to trace the guilty parties. It has, however, been better applied in cases of disasters such as petroleum spills or nuclear incidents where the party responsible for the damage can be easily identified.

Precautionary Principle

A principle in international environmental law and management which states that the lack of conclusive scientific evidence regarding the environmental damage caused by a process should not be used as an excuse to avoid action regarding such an issue-area. The precautionary principle is most often applied in the context of the impact of human actions on the environment and human health, given the complexity of the system and the fact that most of the relationships and cause/effect are extremely difficult to predict in detail. The logic of the precautionary principle is that it is better to err on the side of caution, especially when some of the possible consequences of policies or actions on human health or the natural environment might be irreversible. The precautionary principle is widely accepted in international circles and is frequently mentioned in regional, multilateral and global environmental treaties.

R

Ramsar Convention

The Ramsar Convention is a foremost and well-renowned convention in the area of environmental conservation. Ramsar is indeed one of the oldest global conservation treaties, as the majority of treaties that had been signed in the area of conservation had been bilateral or regional treaties. The convention aims to provide the framework for the conservation of wetlands of international importance. It was signed in Ramsar (now known as Sakht-Sar), Iran, in 1972 and came into force in December 1975. Wetlands are among the world's most productive environments. They are places of enormous biological diversity. They also provide either permanent or temporary homes for large numbers of plant and animal species, including bird, mammal, reptile, amphibian, fish and invertebrate species. Wetlands provide tremendous economic benefits through their role in supporting fisheries, agriculture and tourism, and through much of the world they have a crucial role as a source of clean water for dependent human populations. Unfortunately, they are also among the world's most threatened ecosystems, owing mainly to continued drainage, pollution, over-exploitation or other unsustainable uses of their resources. The convention thus seeks to protect these extremely valuable ecosystems by designating them as conservation areas. In total, there are about 1,675 wetland sites, averaging 150m. ha, designated for inclusion on the Ramsar list of Wetlands of International Importance. Contracting parties are expected to enact national laws to ensure the conservation of these areas or, as a minimum requirement, to replace any original destroyed with one of equal worth. There are presently 155 contracting parties to the convention.

Renewable Energy

The term renewable energy denotes energy derived from ongoing natural processes or resources that are naturally regenerative. Unlike, for example, fossil fuels, of which there is a finite supply, renewable energy is obtained from sources that are essentially inexhaustible as these resources are not depleted in the process of energy generation. Renewable sources of energy include hydroelectricity, wood, waste, wind power, biomass, geothermal, and solar thermal energy. Throughout history mankind has been harnessing energy from a number of these sources in a bid to maintain life on Earth, but it was not until recently that they began to attract wide interest world-wide. The main reason for renewed interest in renewable energy relates first to increased concern over the limit of the non-renewable sources, and second to concern about climate change. Renewables are considered as desirable alternatives because, unlike fossil fuels (coal and petro-leum), the generation of power from renewable sources does not lead to the emission of undesirable gases, such as sulphur oxides (which cause **acid rain**) and greenhouse gases (responsible for climate change). Accordingly, renewable energy is widely seen as a major link in the global transition from a high- to a low-carbon economy. It is for this reason that both industries and governments alike have been committing large sums of money to the research and development of renewable energy over the last two decades or so. These investments have been yielding some positive results. For example, there has been an increase of about 30% in the amount of energy generated through wind sources globally over the last two decades. Indeed, it is now thought that energy from renewable sources accounts for about one-tenth of global energy and this is set to increase over the next few years. In 2005, 4% of the United Kingdom's electricity supply came from renewable sources and the government has set a target of obtaining 10% of electricity supply from renewable energy by 2010. Many other governments across the world and the European Union have also set similar targets to increase the supply of energy from renewable sources. The main challenges involved in generating energy from renewable sources relate to cost, quantity and stability. At the moment, it costs far more to generate energy from non-renewable sources compared with fossil fuels. Besides, it is extremely difficult to derive a high and steady amount of energy from many renewable sources since production is affected by natural variations in meteorological conditions.

S

Silent Spring

Silent Spring is the title of a book written by Rachel Carson, an ecologist and natural historian, published in September 1962 by Houghton Mifflin Press. The book is widely acclaimed as being central among the factors that led to the generation of the new wave of environmental consciousness in the West in the late 1960s and early 1970s. The book contains extensive and graphic documentation of the negative effects of a range of human activities connected with industrialization on the environment. In particular, the book argued that there has been an accumulation of pesticides in the environment and that many dangerous chemicals have found their way into the human food chain. It also claimed that the pesticides were having far-reaching effects on birds. Carson indicated 23 substances belonging to the families of the organochlorine and organophosphate pesticide compounds, such as DDT, the use of which had become commonplace in intensive agriculture. The majority of these substances have since been banned in most industrialized countries. The book was on the top of New York's bestseller list for a long time and has recently been named as one of the greatest science books of all times by the editors of Discovery Magazine.

Sulphur Protocol

The Sulphur Protocol is a protocol negotiated under the **Long Range Transboundary Air Pollution Convention (LTRAP)** in 1994. The protocol was proposed by the UN Economic Commission for Europe (ECE) in order to reduce emissions of sulphur dioxide from power stations and boilers in factories and ships. It was signed and adopted on 14 June in Oslo, Norway, and entered into force on 5 August 1998. The protocol is significant because it is the first international environmental agreement in which parties agreed to set different emission-reduction and clean-up targets for different countries, based on assessments of the 'critical loads' of **acid rain** that ecosystems of forests and lakes can endure before they suffer irreversible damage. The protocol replaced a previous Sulphur Protocol, signed in 1985, under which contracting parties had agreed to reduce SO_x emissions by 30%. Hence, the 1994 agreement is sometimes known as the second sulphur protocol.

Sustainable Development

Sustainable development is a concept that seeks to highlight the need to integrate environmental and social concerns in to mainstream thinking about development. The concept was popularized by the **World Commission on Environment and Development Report** (the Brundtland report) published in 1987. The report defined the concept as development that meets the needs of the present without compromising the ability of future generations to meet their own needs. The concept expresses two key ideas. The first is the need for a departure from the general attitude of regarding the environment as a limitless resource towards seeing it as something that need to be taken into account in mainstream economic planning. Governments, business and industry have traditionally regarded the environment as an almost limitless source of energy and raw materials, with the environmental cost of business regularly shifted to society at large. At the same time, there has traditionally been little thought or discussion on the possible effects of the dominant economic practices of the present generation on the life chances of future generations, in particular the effect on the quality of environment such generations would inherit. The concept thus embodies the notion that development is only sustainable if it allows the present generation to fulfil its life chances and live meaningful lives without unnecessarily polluting the environment and depleting the resource base upon which the development and life chances of future generations depend. This idea is often referred to as the need for intergenerational equity. The second key idea of the concept of sustainable development is global equity and distributional justice. The concept embodies the notion that the world's resources belong to the global community as a whole and that development can only be said to be sustainable if it fulfils the life aspirations not of a few members but of the world community as a whole. It is on this basis that the Brundtland report calls for changes in the world economic structure, and for increased multilateralism and the redistribution of world resources to ensure that the entire global population is given a fair chance of leading meaningful lives. Sustainable development is widely regarded as the main paradigm for the pursuit of economic development across the world. However, there are some who argue that the concept does not have much value since it does not offer a practical guide to policy and economic-planning.

U

UN Environmental Programme (UNEP)

The UN Environmental Programme (UNEP) is a UN agency that evolved from the UN Conference on the Human Environment, held in Stockholm, Sweden, in 1972. UNEP is now accepted as the agency responsible for co-ordinating environmental activities within the UN system. UNEP's global base is in Nairobi, Kenya. It is one of only two UN programmes headquartered in the developing world (the other is UNEP's sister agency UN-HABITAT, which is also located in Nairobi). UNEP neither awards contracts nor executes environmental programmes. Its main responsibility is to stimulate awareness of environmental matters, as well as to provide leadership and encourage partnership in caring for the environment. UNEP executes these functions mainly by inspiring, informing and providing the platform for international co-operation on the environment. Since its inception, UNEP has played a crucial role in the area of issue-definition and agenda-setting. It has functioned as a catalyst for the elaboration of many important environmental regimes, including the **Basel Convention** on the Control of Transboundary Movements of Hazardous Wastes and their Disposal and the Climate Change Convention. In performing its functions, UNEP works closely with other UN agencies and environmental **non-governmental organizations** like the **World Conservation Union,** as well as with environmental scientists, industrialists and decision-makers. UNEP also functions to provide important help to developing countries in their bid to care for the environment. To this end it works with scientists, policy-makers and environmental activists in these countries to promote or strengthen relevant institutions for the wise management of the environment. It also facilitates the transfer of knowledge and technology for **sustainable development** from the more industrialized to the less industrialized countries. UNEP has six regional offices around the world and serves as a host for several environmental convention secretariats, including the Ozone Secretariat, the **Montréal Protocol**'s Multilateral Fund, **CITES** (the Convention on International Trade in Endangered Species of Wild Fauna and Flora), the Convention on Biological Diversity, the Convention on Migratory Species and the Basel Convention.

UN Statement of Principles on Forests

The UN Statement of Principles on Forests refers to the Statement of Principles on the Management, Conservation and **Sustainable Development** of all types of Forests. This is one of the documents agreed at the **Earth Summit** in Rio de Janeiro, Brazil, in 1992. The document, as the name suggests, contains several principles agreed by the global community as providing a good basis for the management of forest lands, both in the developed and the developing countries. The statement of principles is not a legally-binding document as parties are merely encouraged to do their best to apply the principles in the management of forests within their national territories. The statement was essentially a compromise text between the developed and the developing countries which attended the Earth Summit. The developed countries sought a convention that would provide a legally-binding framework for the management of forests on a world-wide basis. The concern for a global convention on forests stems from the rapid, unsustainable rate of **deforestation** in many parts of the world, especially the developing countries. They considered the developing countries were somewhat lax in the way they managed most forests and that more attention is usually paid to the immediate economic benefits that could be derived from deforestation than to the long-term ecological importance of these extremely valuable ecosystems. Many in the developing countries were, however, not convinced that there was a need for such a global convention on forest management, given that most forests lie within the national territories of states. Indeed, for the most part, they thought that the main aim of the developed countries was to gain access to the rich genetic resources in these areas and exert control over what are supposed to be their sovereign resources. For these reasons, they rebuffed the moves for a legally-binding agreement. In the end, governments agreed to adopt the statement, which recognized the need for national policies to ensure the sustainable management of forests, given their vital role in maintaining global ecological processes, but at the same time allowed nations to exploit the resources in the forest in accordance with their specific socio-economic needs. The statement also encourages international co-operation as well as the involvement of indigenous forest dwellers in the management of forests.

V

Vienna Convention for the Protection of the Ozone Layer

The Vienna Convention for the Protection of the Ozone Layer provides the framework for state parties to deal with the problem of ozone-layer destruction. The Vienna Convention was adopted on 22 March 1985 in Vienna, Austria, by 21 states and the European Union, then the European Community. The treaty, which came into force on 22 September 1988, was very effective in raising public awareness of the environmental and health implications of activities resulting or likely to result in the modification or destruction of the ozone layer. It called for a halt in the production of chlorofluorocarbon (CFC) gases and for producers to reduce all emissions of CFC gases. The convention also created a mechanism for intergovernmental co-operation in scientific research, systematic monitoring and exchange of data on the state of the ozone layer, as well as the emission of CFCs and other relevant gases. The convention also laid the foundation for a future protocol that would enable the parties to take more specific actions that might be needed to respond adequately to the issue of ozone-layer destruction.

World Commission on Environment and Development Report (WCED)

Right from the time of its publication in 1986, this text became and has since remained as one of the central texts in global environmental politics. The report was commissioned by the UN General Assembly in 1983 to draw up a global agenda for change mainly by proposing long-term strategies for achieving global **sustainable development** for the year 2000 and beyond. It was authored by a panel of 21 experts, all of whom came from different countries. They held several public meetings across several countries of the world for more than 24 months and took testimonies from people from various professions and walks of life. However, the report eventually assumed the name of the former Norwegian Prime Minister Gro Harlem Brundtland, who chaired the body. It is noted mainly as the work that popularized the term 'sustainable development' and provided detailed arguments, illustrations and statistics to persuade the world that fundamental changes are required in order to secure the well-being and survival of the planet. It is also noted as the report that provided the momentum for the UN Conference on Environment and Development (see **Earth Summit**) in Rio de Janeiro, Brazil, in 1992. The report called for the revival of multilateralism and deliberative democracy at all institutions as a major way of generating consensus on what is needed to ensure that economic development is pursued in a more sustainable manner. Crucially, the report identified inequality as the greatest cause of environmental problems and called on the world's political leaders to work out a system of redistributing resources to the developing countries of the South as a means of achieving global environmental sustainability.

World Conservation Strategy (WCS)

World Conservation Strategy (WCS) is a political document published in 1980 which seeks to articulate a comprehensive view on the notion of conservation. The work also aimed to serve as a resource guide by providing a basic framework for the formulation of national policies on nature and resource conservation. WCS was jointly formulated by the International Union for the Conservation of

Nature and Natural Resources (IUCN) in co-operation with the **UN Environmental Programme (UNEP)**, the World Wide Fund for Nature (WWF), FAO and UNESCO. The document argued for the need to treat the various aspects of the ecosystem as a united entity and the need for governments, businesses and individuals to adopt a long-term approach to economic planning, taking into special consideration the fragile and rapidly degrading nature of the global environmental system. The document warns that the quality of human life is ultimately dependent on the quality of the natural system and cautions against an approach that treats environmental resources as if they were inexhaustible. The document was launched in more than 30 countries and has provided inspiration for many national policies on conservation and **sustainable development**. An updated version, entitled Caring for the Earth: A Strategy for Sustainable Living, was published in 1991.

World Conservation Union (IUCN)

This is the simpler name adopted by the former International Union for Conservation of Nature and Natural Resources in 1990, although it has continued to retain the use of its old acronym as many people still know the organization as IUCN. IUCN is the largest conservation network in the world, with a membership comprising about 83 national governments, 110 governmental organizations and over 800 **non-governmental organizations (NGOs)** and more than 10,000 scientists from all over the world. IUCN is dedicated to influencing and assisting governments and local communities in the pursuit of conservation policies and practices with a view to conserving the integrity and biological diversity of nature throughout the world. The IUCN was founded in October 1948 as the International Union for the Protection of Nature (IUPN). In 1956 it changed its name to the International Union for Conservation of Nature and Natural Resources, and once again in 1990 it changed its name to the World Conservation Union. The IUCN has its headquarters in Gland, Switzerland, and maintains about 1,000 staff located in 40 different countries. The union has played important roles in facilitating the adoption of many international environmental treaties, including the **Ramsar Convention** on Wetlands, the Convention on International Trade in Endangered Species of Wild Flora and Fauna (**CITES**) and the Convention on Biological Diversity. The union engages in extensive research, while also using many of its influential members to create awareness and by lobbying state governments to pursue conservation policies. The union also works with numerous local communities around the world to protect areas and species of natural importance by offering funding, training services and expert scientific advice. The work of the union is critical in maintaining the list of total biological diversity, as well as the list of the endangered species around the world. Some environmental activists and local environmental NGOs accuse the IUCN of being too sensitive to the demands of Western governments while neglecting the needs of the more vulnerable local communities.

The World Conservation Union (IUCN)
28 rue Mauverney,
1196 Gland, Switzerland
Tel: 229990000
Fax: 229990002
Email: webmaster@iucn.org
Internet: www.iucn.org

World Heritage Site

World Heritage Site refers to any natural area of historical sites designated by UNESCO as being of outstanding and universal importance. Countries belonging to the 1972 convention covering the protection of the world's cultural and natural heritage (usually referred to as the World Heritage UNESCO Convention) nominated sites within their countries which they considered worthy of inclusion on the list. To be accepted, however, such a natural site must be: (a) significant in terms of understanding of the Earth's history; (b) a habitat for endangered or rare species; (c) a unique example of natural phenomena; or (d) a place where there is a continuing geological process of important significance. The World Heritage Programme was launched in 1978 and now has a total of about 788 sites in 134 countries included on the list. Some examples include the Great Barrier Reef (Australia), the Galapagos Islands (Ecuador) and the Serengeti National Park (Tanzania).

World Resource Institute (WRI)

The World Resource Institute (WRI) was established in 1982 in Washington, DC, USA, to act as a think-tank on environmental and development issues. The institute was founded with a US $15m. grant from the John D. and Catherine T. MacArthur Foundation of Chicago. Since then it has existed on the basis of donations from government institutions, foundations, corporate organizations, and individuals. The institute prides itself as specializing in high quality, unbiased scientific research and in offering practical solutions that would lead to the solution for global environmental and development issues. The WRI's research agenda covers agriculture, tropical rainforests, climate change, health care and environmental resourcees information systems. The WRI also publishes an annual book of useful environmental data, together with other useful publications in its specific areas of focus.

World Summit on Sustainable Development (WSSD)

The World Summit on Sustainable Development was a global conference held in Johannesburg, South Africa, in August–September 2002 to commemorate the passage of 10 years since the original **Earth Summit** in Rio de Janeiro, Brazil. Although it was well attended, it did not have the same impact or receive the same publicity as the Rio summit. The main aim was to provide a forum to assess the progress made towards achieving global **sustainable development** in line with declarations made at Rio. It was also designed to provide a platform for world leaders to discuss possible ways of meeting the challenges that may have emerged since the first Earth Summit. The WSSD resulted in two main documents: the Johannesburg Declaration on Sustainable Development, which reaffirms the commitment of the world's people to sustainable development; and the Plan of Implementation, which contains specific targets on wide-ranging issues, from poverty eradication to patterns of consumption and production. There was criticism that the conference focused on issues of poverty and failed to make progress on the environment.

Worldwatch Institute

The Worldwatch Institute is a famous non-profit-making research institute based in Washington, DC, USA. The institute was founded by a prominent environmentalist, Lester Brown, in 1974 and focuses on the analysis of trends in major global issues in the area of environment and development. Some of the topics covered by the institute include population, urbanization, oceans, food, water, forests, energy, security and climate change. The institute distributes regular papers to a world-wide audience of decision-makers, scholars and the general public on these issues. It is, however, mostly noted for one of its titles—*The State of the World*—which is an annual publication that provides many graphs, fact-based information and projections on most of the critical issues of the time, especially in the area of development and environment.

Worldwatch Institute
1776 Massachusetts Ave, NW,
Washington, DC 20036-1904, USA
Tel: (202) 452-1999
Fax: (202) 296-7365
Email: worldwatch@worldwatch.org
Internet: www.worldwatch.org

Maps

Maps

Map 1: Protected Areas as a Percentage of Total Land Area

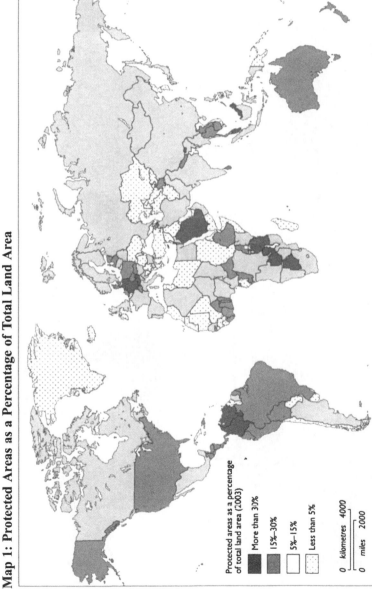

Source data: World Resources Institute 2003. EarthTrends.
Note: Map shows the percentage of total terrestrial area that is listed as 'protected' by the World Database on Protected Areas (WDPA). All terrestrial sites designated by the World Conservation Union (IUCN)—in categories I–VI are included, plus sites recognized by WDPA which have not yet been assigned an IUCN category.

Map 2: World Heritage Sites: Major Wetlands and Marine Sites

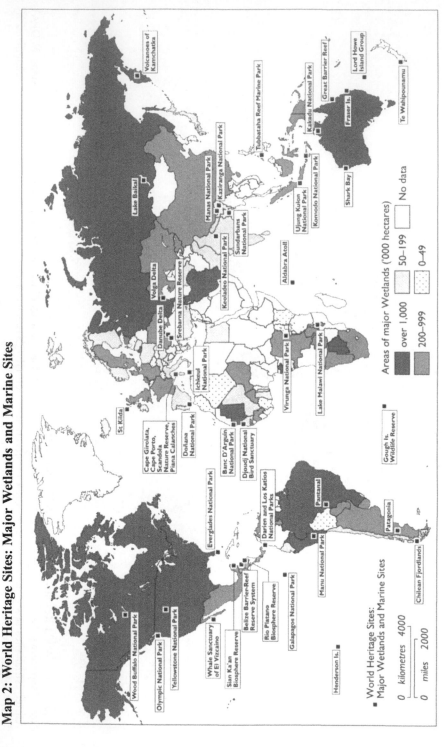

Volcanoes of Kamchatka

Lake Baikal

Manas National Park

Kaziranga National Park

Tubbataha Reef Marine Park

Great Barrier Reef

Lord Howe Island Group

Fraser Is.

Kakadu National Park

Te Wahipounamu

Komodo National Park

Shark Bay

Volga Delta

Sundarbans National Park

Ujung Kulon National Park

Keoladeo National Park

Danube Delta

Srebarna Nature Reserve

Aldabra Atoll

Ichkeul National Park

Virunga National Park

Lake Malawi National Park

St Kilda

Cape Girolata, Cape Porto, Scandola Nature Reserve, Piana Calanches

Doñana National Park

Banc D'Arguin National Park

Djoudj National Bird Sanctuary

Everglades National Park

Gough Is. Wildlife Reserve

Darien and Los Katios National Parks

Pantanal

Patagonia

Whale Sanctuary of El Vizcaino

Sian Ka'an Biosphere Reserve

Belize Barrier-Reef Reserve System

Rio Platano Biosphere Reserve

Manu National Park

Galapagos National Park

Chilean Fjordlands

Wood Buffalo National Park

Olympic National Park

Yellowstone National Park

Henderson Is.

World Heritage Sites:
Major Wetlands and Marine Sites

0   kilometres   4000

0   miles   2000

Areas of major Wetlands ('000 hectares)

over 1,000

200–999

50–199

0–49

No data

Map 3: Areas of Tropical and Other Rainforest Areas

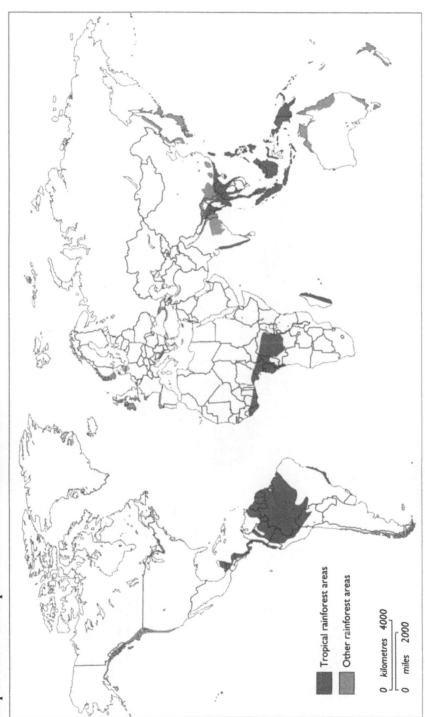

Tropical rainforest areas

Other rainforest areas

0   kilometres   4000

0   miles   2000

Map 4: Hottest and Coldest Climate Regions

Hottest climate regions (Annual average temperature above 20°C)

Coldest climate regions (Annual average temperature below 0°C)

0   kilometres   4000

0   miles   2000

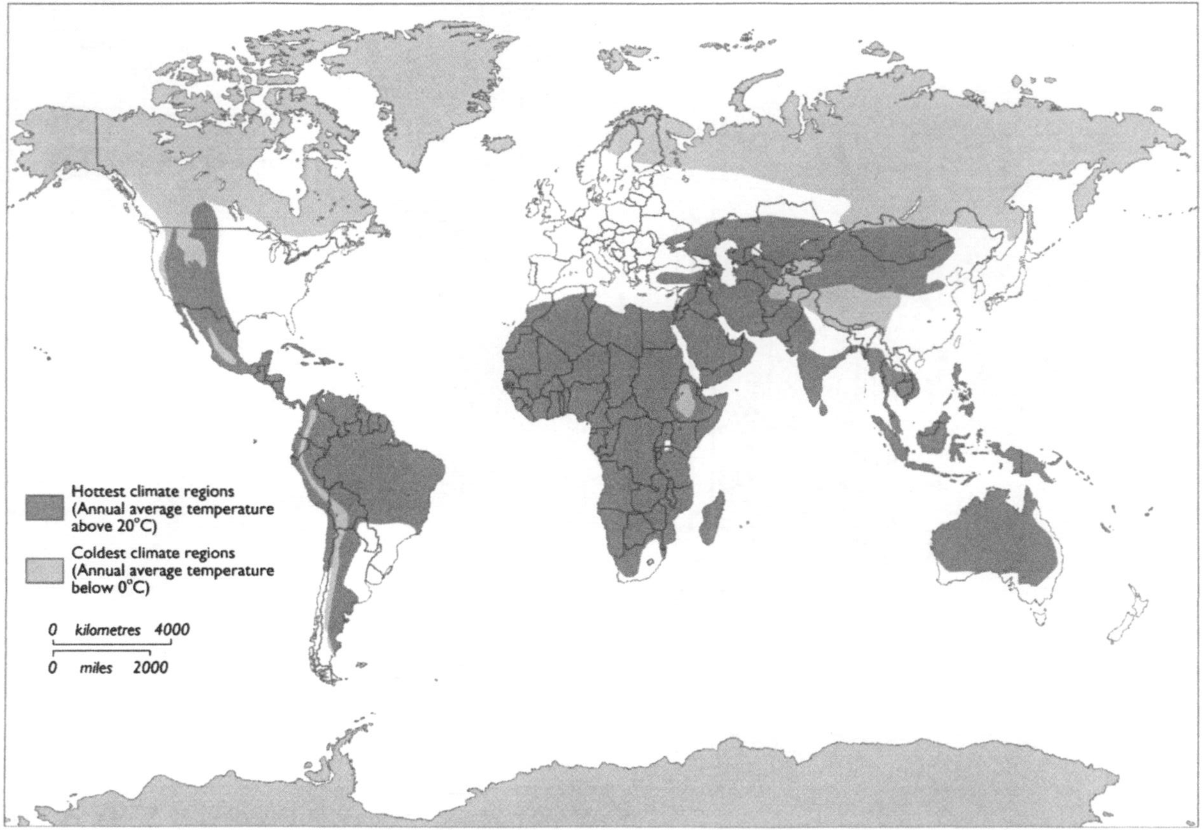

Map 5: Total Annual Carbon Dioxide (CO₂) Emissions

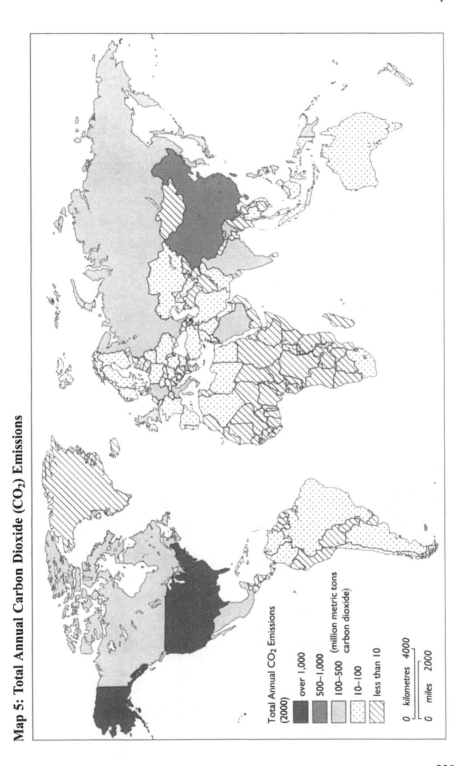

Total Annual CO₂ Emissions
(2000)

over 1,000
500–1,000
100–500 (million metric tons
10–100   carbon dioxide)
less than 10

0 kilometres 4000
0 miles 2000

Map 6: Annual Energy Consumption per Caput

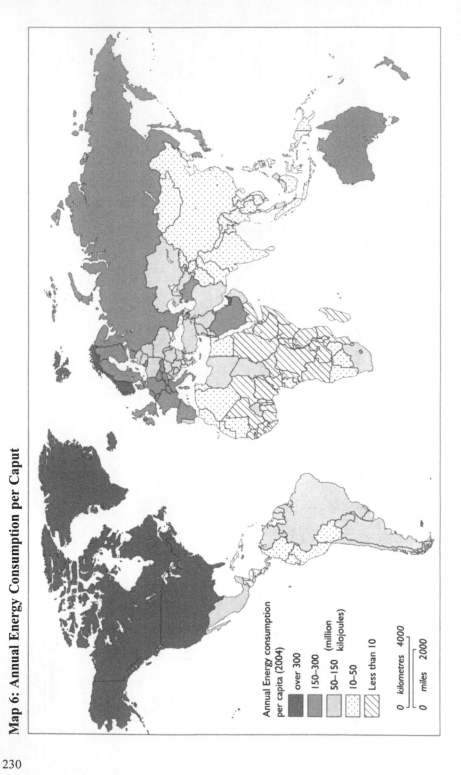

Annual Energy consumption
per capita (2004)

over 300

150–300

50–150   (million
            kilojoules)

10–50

Less than 10

0   kilometres 4000

0   miles   2000

Map 7: Deserts and Desertification

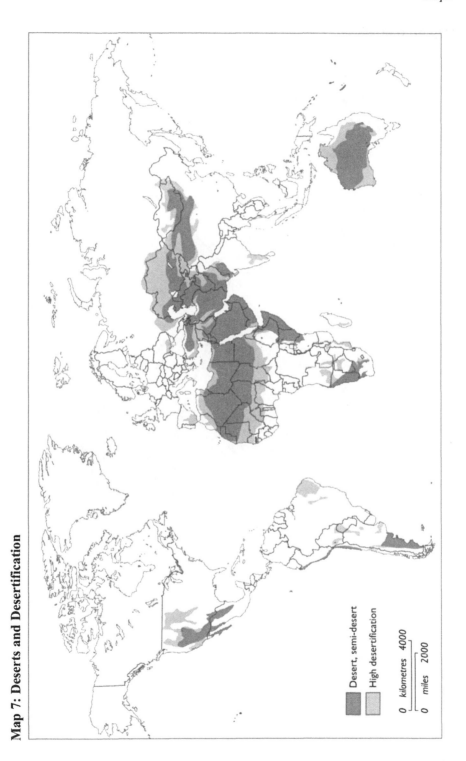

Desert, semi-desert

High desertification

kilometres

miles

Map 8: Antarctica

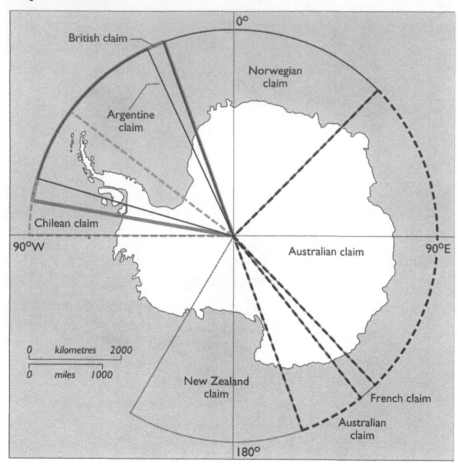

Statistics

Table 1: Demographic trends—total population (millions)
(by Human Development Index Rank)

Refers to the *de facto* population, which includes all people actually present in a given area at a given time.

HDI Rank*	Country	1975	2004	2015–[1]
1	Norway	4.0	4.6	4.8
2	Iceland	0.2	0.3	0.3
3	Australia	13.6	19.9	22.2
4	Ireland	3.2	4.1	4.7
5	Sweden	8.2	9.0	9.3
6	Canada	23.1	32.0	35.1
7	Japan	111.5	127.9	128.0
8	USA	220.2	295.4	325.7
9	Switzerland	6.3	7.2	7.3
10	Netherlands	13.7	16.2	16.8
11	Finland	4.7	5.2	5.4
12	Luxembourg	0.4	0.5	0.5
13	Belgium	9.8	10.4	10.5
14	Austria	7.6	8.2	8.3
15	Denmark	5.1	5.4	5.6
16	France	52.7	60.3	62.3
17	Italy	55.4	58.0	57.8
18	United Kingdom	55.4	59.5	61.4
19	Spain	35.6	42.6	44.4
20	New Zealand	3.1	4.0	4.3
21	Germany	78.7	82.6	82.5
22	Hong Kong, China (SAR)	4.4	7.0	7.8
23	Israel	3.4	6.6	7.8
24	Greece	9.0	11.1	11.2
25	Singapore	2.3	4.3	4.8
26	Korea, Rep. of	35.3	47.6	49.1
27	Slovenia	1.7	2.0	1.9
28	Portugal	9.1	10.4	10.8
29	Cyprus	0.6	0.8	0.9
30	Czech Republic	10.0	10.2	10.1
31	Barbados	0.2	0.3	0.3
32	Malta	0.3	0.4	0.4
33	Kuwait	1.0	2.6	3.4
34	Brunei Darussalam	0.2	0.4	0.5
35	Hungary	10.5	10.1	9.8
36	Argentina	26.0	38.4	42.7
37	Poland	34.0	38.6	38.1
38	Chile	10.4	16.1	17.9

Continued

HDI Rank*	Country	1975	2004	2015–[1]
39	Bahrain	0.3	0.7	0.9
40	Estonia	1.4	1.3	1.3
41	Lithuania	3.3	3.4	3.3
42	Slovakia	4.7	5.4	5.4
43	Uruguay	2.8	3.4	3.7
44	Croatia	4.3	4.5	4.5
45	Latvia	2.5	2.3	2.2
46	Qatar	0.2	0.8	1.0
47	Seychelles	0.1	0.1	0.1
48	Costa Rica	2.1	4.3	5.0
49	United Arab Emirates	0.5	4.3	5.6
50	Cuba	9.3	11.2	11.4
51	Saint Christopher and Nevis[2]	...	...	...
52	Bahamas	0.2	0.3	0.4
53	Mexico	59.3	105.7	119.1
54	Bulgaria	8.7	7.8	7.2
55	Tonga	0.1	0.1	0.1
56	Oman	0.9	2.5	3.2
57	Trinidad and Tobago	1.0	1.3	1.3
58	Panama	1.7	3.2	3.8
59	Antigua and Barbuda	0.1	0.1	0.1
60	Romania	21.2	21.8	20.9
61	Malaysia	12.3	24.9	29.6
62	Bosnia and Herzegovina	3.7	3.9	3.9
63	Mauritius	0.9	1.2	1.3
64	Libyan Arab Jamahiriya	2.4	5.7	7.0
65	Russian Federation	134.2	143.9	136.7
66	Macedonia, FYR	1.7	2.0	2.1
67	Belarus	9.4	9.8	9.2
68	Dominica	0.1	0.1	0.1
69	Brazil	108.1	183.9	209.4
70	Colombia	25.4	44.9	52.1
71	Saint Lucia	0.1	0.2	0.2
72	Venezuela, RB	12.7	26.3	31.3
73	Albania	2.4	3.1	3.3
74	Thailand	41.3	63.7	69.1
75	Samoa (Western)	0.2	0.2	0.2
76	Saudi Arabia	7.3	24.0	30.8
77	Ukraine	49.0	47.0	41.8
78	Lebanon	2.7	3.5	4.0
79	Kazakhstan	14.1	14.8	14.9
80	Armenia	2.8	3.0	3.0

Continued

HDI Rank*	Country	1975	2004	2015–[1]
81	China	927.8³	1,308.0³	1,393.0³
82	Peru	15.2	27.6	32.2
83	Ecuador	6.9	13.0	15.1
84	Philippines	42.0	81.6	96.8
85	Grenada	0.1	0.1	0.1
86	Jordan	1.9	5.6	7.0
87	Tunisia	5.7	10.0	11.1
88	Saint Vincent and the Grenadines	0.1	0.1	0.1
89	Suriname	0.4	0.4	0.5
90	Fiji	0.6	0.8	0.9
91	Paraguay	2.7	6.0	7.6
92	Turkey	41.2	72.2	82.6
93	Sri Lanka	14.0	20.6	22.3
94	Dominican Republic	5.1	8.8	10.1
95	Belize	0.1	0.3	0.3
96	Iran, Islamic Rep. of	33.3	68.8	79.9
97	Georgia	4.9	4.5	4.2
98	Maldives	0.1	0.3	0.4
99	Azerbaijan	5.7	8.4	9.1
100	Occupied Palestinian Territories	1.3	3.6	5.0
101	El Salvador	4.1	6.8	8.0
102	Algeria	16.0	32.4	38.1
103	Guyana	0.7	0.8	0.7
104	Jamaica	2.0	2.6	2.7
105	Turkmenistan	2.5	4.8	5.5
106	Cape Verde	0.3	0.5	0.6
107	Syrian Arab Republic	7.5	18.6	23.8
108	Indonesia	134.4	220.1	246.8
109	Viet Nam	48.0	83.1	95.0
110	Kyrgyzstan	3.3	5.2	5.9
111	Egypt	39.3	72.6	88.2
112	Nicaragua	2.6	5.4	6.6
113	Uzbekistan	14.0	26.2	30.7
114	Moldova, Rep. of	3.8	4.2	4.1
115	Bolivia	4.8	9.0	10.9
116	Mongolia	1.4	2.6	3.0
117	Honduras	3.0	7.0	8.8
118	Guatemala	6.2	12.3	15.9
119	Vanuatu	0.1	0.2	0.3
120	Equatorial Guinea	0.2	0.5	0.6

Continued

HDI Rank*	Country	1975	2004	2015–[1]
121	South Africa	25.9	47.2	47.9
122	Tajikistan	3.4	6.4	7.6
123	Morocco	17.3	31.0	36.2
124	Gabon	0.6	1.4	1.6
125	Namibia	0.9	2.0	2.2
126	India	620.7	1,087.1	1,260.4
127	São Tomé and Príncipe	0.1	0.2	0.2
128	Solomon Islands	0.2	0.5	0.6
129	Cambodia	7.1	13.8	17.1
130	Myanmar	30.1	50.0	55.0
131	Botswana	0.9	1.8	1.7
132	Comoros	0.3	0.8	1.0
133	Lao People's Dem. Rep.	3.0	5.8	7.3
134	Pakistan	68.3	154.8	193.4
135	Bhutan	1.2	2.1	2.7
136	Ghana	10.2	21.7	26.6
137	Bangladesh	73.2	139.2	168.2
138	Nepal	13.5	26.6	32.7
139	Papua New Guinea	2.9	5.8	7.0
140	Congo, Rep. of the	1.5	3.9	5.4
141	Sudan	17.1	35.5	44.0
142	Timor-Leste	0.7	0.9	1.5
143	Madagascar	7.9	18.1	23.8
144	Cameroon	7.6	16.0	19.0
145	Uganda	10.8	27.8	41.9
146	Swaziland	0.5	1.0	1.0
147	Togo	2.4	6.0	7.8
148	Djibouti	0.2	0.8	0.9
149	Lesotho	1.1	1.8	1.7
150	Yemen	7.0	20.3	28.5
151	Zimbabwe	6.2	12.9	13.8
152	Kenya	13.5	33.5	44.2
153	Mauritania	1.4	3.0	4.0
154	Haiti	4.9	8.4	9.8
155	Gambia	0.6	1.5	1.9
156	Senegal	5.3	11.4	14.5
157	Eritrea	2.1	4.2	5.8
158	Rwanda	4.4	8.9	11.3
159	Nigeria	58.9	128.7	160.9
160	Guinea	4.2	9.2	11.9
161	Angola	6.8	15.5	20.9
162	Tanzania	16.0	37.6	45.6

Continued

HDI Rank*	Country	1975	2004	2015–[1]
163	Benin	3.2	8.2	11.2
164	Côte d'Ivoire	6.6	17.9	21.6
165	Zambia	5.2	11.5	13.8
166	Malawi	5.2	12.6	16.0
167	Congo, Dem. Rep. of the	23.9	55.9	78.0
168	Mozambique	10.6	19.4	23.5
169	Burundi	3.7	7.3	10.6
170	Ethiopia	34.1	75.6	97.2
171	Chad	4.2	9.4	12.8
172	Central African Republic	2.1	4.0	4.6
173	Guinea-Bissau	0.7	1.5	2.1
174	Burkina Faso	5.9	12.8	17.7
175	Mali	6.2	13.1	18.1
176	Sierra Leone	2.9	5.3	6.9
177	Niger	5.3	13.5	19.3

Notes: * The HDI (human development index) is a summary composite index that measures a country's average achievements in three basic aspects of human development: health, knowledge, and a decent standard of living. Health is measured by life expectancy at birth; knowledge is measured by a combination of the adult literacy rate and the combined primary, secondary, and tertiary education gross enrolment ratio; and standard of living by GDP per caput (PPP US$).

1 – Data refer to medium-variant projections.

2 – Total population in 2004 was 42,000.

3 – Population estimates include Taiwan, province of China.

Source: Human Development Report 2006, United Nations Development Programme. UN (United Nations). 2005b. *World Population Prospects 1950–2050: The 2004 Revision.* Database. Department of Economic and Social Affairs, Population Division. New York. Reproduced with permission of Palgrave Macmillan.

Table 2: Demographic trends—annual population growth rate (%)*
(by Human Development Index Rank)

HDI Rank	Country	1975–2004	2004–15[1]
1	Norway	0.5	0.5
2	Iceland	1.0	0.8
3	Australia	1.3	1.0
4	Ireland	0.9	1.2
5	Sweden	0.3	0.3
6	Canada	1.1	0.8
7	Japan	0.5	...
8	USA	1.0	0.9
9	Switzerland	0.5	0.1
10	Netherlands	0.6	0.3
11	Finland	0.4	0.2
12	Luxembourg	0.8	1.2
13	Belgium	0.2	0.1
14	Austria	0.3	0.1
15	Denmark	0.2	0.2
16	France	0.5	0.3
17	Italy	0.2	...
18	United Kingdom	0.2	0.3
19	Spain	0.6	0.4
20	New Zealand	0.9	0.7
21	Germany	0.2	...
22	Hong Kong, China (SAR)	1.6	1.0
23	Israel	2.3	1.6
24	Greece	0.7	0.1
25	Singapore	2.2	1.1
26	Korea, Rep. of	1.0	0.3
27	Slovenia	0.4	−0.1
28	Portugal	0.5	0.3
29	Cyprus	1.0	1.0
30	Czech Republic	0.1	−0.1
31	Barbados	0.3	0.2
32	Malta	0.9	0.4
33	Kuwait	3.3	2.4
34	Brunei Darussalam	2.8	2.0
35	Hungary	−0.1	−0.3
36	Argentina	1.3	1.0
37	Poland	0.4	−0.1
38	Chile	1.5	1.0
39	Bahrain	3.3	1.6
40	Estonia	−0.2	−0.3

Continued

HDI Rank	Country	1975–2004	2004–15[1]
41	Lithuania	0.1	−0.4
42	Slovakia	0.5	...
43	Uruguay	0.7	0.6
44	Croatia	0.2	−0.2
45	Latvia	−0.2	−0.5
46	Qatar	5.2	2.0
47	Seychelles	1.0	0.9
48	Costa Rica	2.5	1.4
49	United Arab Emirates	7.2	2.4
50	Cuba	0.7	0.2
51	Saint Christopher and Nevis	−0.2	1.1
52	Bahamas	1.8	1.2
53	Mexico	2.0	1.1
54	Bulgaria	−0.4	−0.8
55	Tonga	0.4	0.1
56	Oman	3.5	2.0
57	Trinidad and Tobago	0.9	0.3
58	Panama	2.1	1.6
59	Antigua and Barbuda	0.9	1.2
60	Romania	0.1	−0.4
61	Malaysia	2.4	1.6
62	Bosnia and Herzegovina	0.1	...
63	Mauritius	1.1	0.8
64	Libyan Arab Jamahiriya	2.9	1.8
65	Russian Federation	0.2	−0.5
66	Macedonia, FYR	0.7	0.1
67	Belarus	0.2	−0.6
68	Dominica	0.3	0.9
69	Brazil	1.8	1.2
70	Colombia	2.0	1.3
71	Saint Lucia	1.3	0.8
72	Venezuela	2.5	1.6
73	Albania	0.9	0.6
74	Thailand	1.5	0.7
75	Samoa (Western)	0.7	0.3
76	Saudi Arabia	4.1	2.3
77	Ukraine	−0.1	−1.1
78	Lebanon	1.0	1.0
79	Kazakhstan	0.2	...
80	Armenia	0.2	−0.2
81	China	1.2[2]	0.6[2]
82	Peru	2.1	1.4

Continued

HDI Rank	Country	1975–2004	2004–15[1]
83	Ecuador	2.2	1.4
84	Philippines	2.3	1.6
85	Grenada	0.4	1.3
86	Jordan	3.6	2.0
87	Tunisia	2.0	1.0
88	Saint Vincent and the Grenadines	0.7	0.4
89	Suriname	0.7	0.5
90	Fiji	1.3	0.6
91	Paraguay	2.8	2.1
92	Turkey	1.9	1.2
93	Sri Lanka	1.3	0.7
94	Dominican Republic	1.9	1.3
95	Belize	2.3	1.8
96	Iran, Islamic Rep. of	2.5	1.4
97	Georgia	−0.3	−0.7
98	Maldives	2.9	2.4
99	Azerbaijan	1.3	0.8
100	Occupied Palestinian Territories	3.6	3.0
101	El Salvador	1.7	1.5
102	Algeria	2.4	1.5
103	Guyana	0.1	−0.1
104	Jamaica	0.9	0.4
105	Turkmenistan	2.2	1.3
106	Cape Verde	2.0	2.2
107	Syrian Arab Republic	3.1	2.3
108	Indonesia	1.7	1.0
109	Viet Nam	1.9	1.2
110	Kyrgyzstan	1.6	1.1
111	Egypt	2.1	1.8
112	Nicaragua	2.5	1.9
113	Uzbekistan	2.2	1.4
114	Moldova, Rep. of	0.3	−0.2
115	Bolivia	2.2	1.7
116	Mongolia	2.0	1.2
117	Honduras	2.9	2.0
118	Guatemala	2.4	2.3
119	Vanuatu	2.5	1.8
120	Equatorial Guinea	2.7	2.2
121	South Africa	2.1	0.1
122	Tajikistan	2.2	1.5
123	Morocco	2.0	1.4
124	Gabon	2.8	1.5

Continued HDI Rank	Country	1975–2004	2004–15[1]
125	Namibia	2.8	1.0
126	India	1.9	1.3
127	São Tomé and Príncipe	2.1	2.1
128	Solomon Islands	3.0	2.2
129	Cambodia	2.3	1.9
130	Myanmar	1.7	0.9
131	Botswana	2.4	−0.4
132	Comoros	3.1	2.5
133	Lao People's Dem. Rep.	2.2	2.1
134	Pakistan	2.8	2.0
135	Bhutan	2.1	2.2
136	Ghana	2.6	1.9
137	Bangladesh	2.2	1.7
138	Nepal	2.3	1.9
139	Papua New Guinea	2.4	1.8
140	Congo, Rep. of the	3.2	3.1
141	Sudan	2.5	2.0
142	Timor-Leste	1.0	4.7
143	Madagascar	2.9	2.5
144	Cameroon	2.6	1.6
145	Uganda	3.3	3.7
146	Swaziland	2.3	−0.4
147	Togo	3.1	2.5
148	Djibouti	4.3	1.6
149	Lesotho	1.6	−0.3
150	Yemen	3.7	3.1
151	Zimbabwe	2.5	0.6
152	Kenya	3.1	2.5
153	Mauritania	2.5	2.6
154	Haiti	1.8	1.3
155	Gambia	3.4	2.2
156	Senegal	2.7	2.2
157	Eritrea	2.4	2.9
158	Rwanda	2.4	2.2
159	Nigeria	2.7	2.0
160	Guinea	2.7	2.3
161	Angola	2.8	2.7
162	Tanzania	2.9	1.7
163	Benin	3.2	2.9
164	Côte d'Ivoire	3.4	1.7
165	Zambia	2.8	1.7
166	Malawi	3.0	2.2

Continued

HDI Rank	Country	1975–2004	2004–15[1]
167	Congo, Dem. Rep. of the	2.9	3.0
168	Mozambique	2.1	1.7
169	Burundi	2.4	3.4
170	Ethiopia	2.7	2.3
171	Chad	2.8	2.8
172	Central African Republic	2.3	1.4
173	Guinea-Bissau	3.0	3.0
174	Burkina Faso	2.6	2.9
175	Mali	2.6	2.9
176	Sierra Leone	2.1	2.3
177	Niger	3.2	3.2

Notes:

* Refers to the average annual exponential growth rate for the period indicated.

1 – Data refer to medium-variant projections.

2 – Population estimates include Taiwan, province of China.

Source: Human Development Report 2006, United Nations Development Programme.
Reproduced with permission of Palgrave Macmillan.

Table 3: Energy and resources—energy consumption: total energy consumption per caput

(Kilograms of oil equivalent (kgoe) per person)

Country	1990	2000	2003
Albania	809.30	550.30	673.50
Algeria	943.30	960.20	1,037.70
Angola	596.70	572.30	606.10
Argentina	1,415.30	1,678.80	1,574.80
Armenia	...	672.40	659.70
Australia	5,188.50	5,793.40	5,723.30
Austria	3,238.50	3,578.50	4,053.20
Azerbaijan	...	1,413.70	1,480.30
Bahrain	9,795.70	9,277.80	10,250.50
Bangladesh	123.30	145.10	160.90
Belarus	...	2,453.80	2,630.90
Belgium	4,927.10	5,714.80	5,703.40
Benin	324.10	279.50	301.40
Bolivia	415.90	594.30	503.80
Bosnia and Herzegovina	...	1,044.40	1,135.10
Botswana	889.90	1,050.20	1,049.20
Brazil	896.60	1,068.10	1,067.60
Brunei Darussalam	7,070.20	7,632.80	7,485.10
Bulgaria	3,305.80	2,337.90	2,508.00
Cameroon	431.90	428.70	434.10
Canada	7,558.40	8,151.30	8,300.70
Côte d'Ivoire	348.30	409.70	378.80
Chile	1,067.30	1,684.10	1,652.20
China[1]	791.70	946.40	1,138.30
Colombia	716.30	672.00	636.90
Congo, Rep. of the	425.00	246.80	272.70
Congo, Dem. Rep. of the	315.20	294.90	296.20
Costa Rica	658.40	842.00	879.90
Croatia	...	1,725.60	1,941.50
Cuba	1,594.10	1,031.90	935.10
Cyprus	2,256.10	3,085.10	3,281.10
Czech Republic	4,753.80	3,934.50	4,319.30
Denmark	3,480.50	3,620.10	3,832.80
Dominican Republic	583.80	944.10	922.40
Ecuador	596.60	679.00	781.50
Egypt	572.90	681.40	761.30
El Salvador	496.10	649.10	683.20
Eritrea	...	202.60	199.30
Estonia	...	3,313.00	3,672.40

Continued

Country	1990	2000	2003
Ethiopia[2]	280.20	273.20	277.90
Finland	5,850.50	6,373.10	7,218.10
France	4,005.90	4,345.10	4,518.40
Gabon	1,299.20	1,212.40	1,248.70
Georgia	...	613.40	600.70
Germany	4,484.50	4,173.00	4,203.10
Ghana	344.80	397.00	400.20
Gibraltar	2,464.40	4,774.60	5,104.60
Greece	2,183.20	2,535.00	2,698.60
Guatemala	503.50	640.00	607.90
Haiti	230.80	256.90	270.00
Honduras	496.50	468.90	521.90
Hong Kong	1,869.40	2,328.20	2,398.90
Hungary	2,755.20	2,446.10	2,595.20
Iceland	8,515.70	11,545.00	11,718.00
India	425.70	501.40	512.40
Indonesia	538.00	697.50	757.40
Iran, Islamic Rep. of	1,213.50	1,787.80	2,034.10
Iraq	1,029.40	1,036.80	950.60
Ireland	2,961.20	3,758.20	3,761.30
Israel	2,683.10	3,160.00	3,187.90
Italy	2,609.90	2,992.80	3,127.20
Jamaica	1,242.40	1,516.60	1,545.10
Japan	3,610.00	4,163.70	4,040.40
Jordan	1,074.90	1,043.70	1,022.40
Kazakhstan	...	2,640.00	3,359.00
Kenya	532.60	490.00	481.20
Korea, Dem. People's Rep.	1,669.60	903.30	894.10
Korea, Rep. of	2,161.20	4,080.70	4,346.50
Kuwait	3,951.10	9,174.20	9,076.00
Kyrgyzstan	...	493.70	520.50
Latvia	...	1,638.20	1,888.70
Lebanon	842.60	1,488.50	1,700.10
Libyan Arab Jamahiriya	2,662.80	3,207.70	3,203.20
Lithuania	...	2,086.30	2,629.20
Luxembourg	9,448.30	8,467.00	9,408.80
Macedonia, FYR	...	1,348.10	1,313.90
Malaysia	1,268.50	2,122.80	2,318.40
Malta	2,150.90	2,002.00	2,242.00
Mexico	1,475.00	1,502.40	1,533.20
Moldova, Rep. of	...	671.50	787.50
Morocco	272.30	341.20	357.30

Continued

Country	1990	2000	2003
Mozambique	536.40	402.20	435.80
Myanmar	262.10	265.20	276.50
Namibia	...	544.20	640.00
Nepal	303.80	334.20	335.90
Netherlands	4,463.70	4,769.70	5,012.20
Netherlands Antilles	7,819.20	7,893.00	9,198.50
New Zealand	4,033.40	4,542.60	4,378.60
Nicaragua	535.20	554.20	593.80
Nigeria	783.00	759.10	776.90
Norway	5,067.70	5,729.10	5,933.60
Oman	2,475.50	3,959.70	4,975.00
Pakistan	388.80	448.30	456.70
Panama	618.00	874.90	835.90
Paraguay	730.80	706.40	678.70
Peru	457.50	482.80	431.50
Philippines	428.10	559.80	524.90
Poland	2,620.60	2,312.90	2,369.70
Portugal	1,777.60	2,473.30	2,482.00
Qatar	14,788.80	21,439.80	21,395.80
Romania	2,689.00	1,640.30	1,784.00
Russian Federation	...	4,189.10	4,423.20
Saudi Arabia	4,113.80	5,272.10	5,582.20
Senegal	280.50	239.40	233.20
Serbia and Montenegro	1,730.80	1,265.60	1,538.80
Singapore	4,428.70	5,538.60	5,158.70
Slovakia	4,055.40	3,284.80	3,448.60
Slovenia	...	3,298.80	3,561.70
South Africa	2,473.90	2,475.20	2,596.90
Spain	2,317.20	3,062.00	3,228.40
Sri Lanka	310.10	407.20	423.80
Sudan	408.30	419.30	475.90
Sweden	5,557.50	5,487.50	5,764.80
Switzerland	3,657.00	3,649.90	3,718.60
Syrian Arab Republic	909.20	1,044.70	981.70
Tajikistan	...	470.70	501.20
Tanzania	373.90	386.90	464.90
Thailand	802.70	1,213.70	1,405.70
Togo	365.40	386.50	445.30
Trinidad and Tobago	4,968.50	7,670.30	8,555.10
Tunisia	673.60	793.70	833.30
Turkey	924.40	1,128.90	1,105.80
Turkmenistan	...	3,214.70	3,646.40

Continued

Country	1990	2000	2003
Ukraine	...	2,646.90	2,968.00
United Arab Emirates	12,069.00	11,022.90	10,538.70
United Kingdom	3,738.10	3,970.20	3,918.10
USA	7,543.40	8,109.00	7,794.80
Uruguay	724.80	920.90	737.10
Uzbekistan	...	2,028.50	2,043.20
Venezuela	2,225.40	2,323.10	2,057.00
Viet Nam	367.40	475.70	539.40
Yemen	212.30	271.70	294.80
Zambia	653.00	586.10	600.60
Zimbabwe	888.20	795.70	743.80
World	1,633.30	1,633.80	1,674.40

1. Figures for China Include totals for Taiwan.

2. Ethiopia: 1990 value includes totals for Eritrea.

Source: World Resources Institute. 2007. EarthTrends: Environmental Information. Available at http://earthtrends.wri.org. Washington, DC: World Resources Institute.

International Energy Agency (IEA) Statistics Division. 2006. *Energy Balances of OECD Countries* (2006 edition) and *Energy Balances of Non-OECD Countries* (2006 edition). Paris: IEA. Available at http://data.iea.org/ieastore/default.asp.

Population Division of the Department of Economic and Social Affairs of the United Nations Secretariat. 2005. *World Population Prospects: The 2004 Revision*. Dataset on CD-ROM. New York: United Nations. Available at http://www.un.org/esa/population/publications/WPP2004/wpp2004.htm.

Technical Notes: Total energy consumption per caput measures the amount of primary energy consumed, on average, by each person living in a particular country or region for the year indicated. All primary sources of energy, including coal and coal products, oil and petroleum products, natural gas, nuclear, hydroelectric, etc., are included here. Data are reported in kilograms of oil equivalent (kgoe) per person.

Consumption equals indigenous production + imports – exports – energy delivered to international marine bunkers +/– stock changes. The International Energy Agency (IEA) refers to these data as Total Primary Energy Supply (TPES). Energy losses from transportation, friction, heat, and other inefficiencies are included in these totals.

Methodology: World Resources Institute (WRI) calculates per caput energy consumption with population data from the United Nations Population Division.

These statistics are expressed in terms of 'net' calorific value, so the values reported here may be slightly lower than those in other statistical compendia which report energy in terms of 'gross' calorific value.

Table 4: Greenhouse gas emissions—CO_2 emissions (2003)

	CO_2 emissions	% change since 1990	CO_2 emissions per caput	CO_2 emissions per km^2
	m. tons	*%*	*ton / person*	*ton / km^2*
Afghanistan	0.70	−73.10	0.03	1.08
Albania	3.05	−58.20	0.98	105.92
Algeria	163.95	112.60	5.14	68.83
American Samoa	0.29	2.10	4.72	1,467.34
Angola	8.63	85.50	0.57	6.93
Antigua and Barbuda	0.40	32.60	5.01	902.71
Argentina	127.73	16.20	3.36	45.94
Armenia	3.43	...	1.13	115.17
Aruba	2.16	17.20	22.30	11,983.33
Australia	371.70	32.30	18.80	48.02
Austria	76.21	24.40	9.40	908.80
Azerbaijan	29.22	...	3.52	337.45
Bahamas	1.87	−4.00	5.96	134.96
Bahrain	21.91	86.80	31.04	31,573.49
Bangladesh	34.69	125.50	0.25	240.91
Barbados	1.19	10.70	4.44	2,772.09
Belarus	52.59	−48.60	5.30	253.32
Belgium	126.20	6.00	12.20	4,133.91
Belize	0.78	149.20	3.01	33.96
Benin	2.05	186.00	0.26	18.16
Bermuda	0.50	−15.60	7.82	9,396.23
Bhutan	0.39	202.30	0.19	8.23
Bolivia	7.91	43.60	0.90	7.20
Bosnia and Herzegovina	19.16	...	4.89	374.26
Botswana	4.12	89.70	2.33	7.09
Brazil	298.90	47.30	1.65	35.10
British Virgin Islands	0.08	57.10	3.59	509.93
Brunei Darussalam	4.56	−21.80	12.75	790.63
Bulgaria	53.32	−45.90	6.80	480.74
Burkina Faso	1.04	4.50	0.08	3.80
Burundi	0.24	22.30	0.03	8.48
Cambodia	0.54	18.40	0.04	2.96
Cameroon	3.54	120.30	0.22	7.45
Canada	586.07	27.50	18.50	58.78
Cape Verde	0.14	73.50	0.30	35.71
Cayman Islands	0.30	22.10	7.05	1,151.52
Central African Republic	0.25	26.00	0.06	0.40
Chad	0.12	−18.70	0.01	0.09
Chile	58.59	65.60	3.67	77.49
China	4,151.41	72.80	3.19	432.58
Colombia	55.63	−2.20	1.26	48.85
Comoros	0.09	36.90	0.12	39.82

Continued

	CO$_2$ emissions	% change since 1990	CO$_2$ emissions per caput	CO$_2$ emissions per km^2
	m. tons	*%*	*ton / person*	*ton / km^2*
Congo, Republic of the	1.38	17.60	0.37	4.04
Cook Islands	0.03	40.90	1.70	131.36
Costa Rica	6.34	117.00	1.52	124.07
Côte d'Ivoire	5.72	6.10	0.33	17.75
Croatia	23.00	−0.20	5.10	406.81
Cuba	25.30	−21.20	2.25	228.17
Cyprus	7.29	56.60	8.93	788.13
Czech Republic	127.12	−22.50	12.40	1611.85
Congo, Dem. Rep. of the	1.79	−55.00	0.03	0.76
Denmark	60.75	12.00	11.30	1,409.71
Djibouti	0.37	3.70	0.48	15.78
Dominica	0.14	137.90	1.76	183.75
Dominican Republic	21.35	122.90	2.47	438.60
Ecuador	23.25	40.10	1.81	81.98
Egypt	139.89	85.10	1.96	139.69
El Salvador	6.55	150.00	0.99	311.44
Equatorial Guinea	0.17	41.90	0.34	5.92
Eritrea	0.70	...	0.17	5.97
Estonia	19.11	−49.90	14.20	423.73
Ethiopia	7.35	147.50	0.10	6.65
Faeroe Islands	0.66	7.00	14.22	474.52
Falkland Islands (Malvinas)	0.05	21.10	15.06	3.78
Fiji	1.12	37.30	1.34	61.29
Finland	73.19	30.00	14.00	216.45
France	408.16	2.80	6.80	740.09
French Guiana	1.00	24.80	5.64	11.17
French Polynesia	0.69	13.60	2.79	173.50
Gabon	1.23	−79.60	0.91	4.58
Gambia	0.28	48.20	0.20	25.06
Georgia	3.73	...	0.82	53.54
Germany	865.37	−14.70	10.50	2,423.86
Ghana	7.74	105.40	0.37	32.47
Gibraltar	0.36	495.10	13.04	60500.00
Greece	109.98	30.90	9.90	833.45
Greenland	0.57	2.70	10.03	0.26
Grenada	0.22	84.20	2.17	642.44
Guadeloupe	1.71	33.50	3.88	1,004.69
Guam	4.09	80.00	24.95	7,444.44
Guatemala	10.71	110.30	0.89	98.37
Guinea	1.34	32.20	0.15	5.45
Guinea-Bissau	0.27	29.20	0.18	7.47
Guyana	1.63	43.90	2.18	7.59

Continued

	CO_2 emissions	% change since 1990	CO_2 emissions per caput	CO_2 emissions per km^2
	m. tons	*%*	*ton / person*	*ton / km^2*
Haiti	1.74	75.00	0.21	62.74
Honduras	6.51	150.90	0.94	58.05
Hong Kong, China (SAR)	37.87	44.40	5.50	34,454.05
Hungary	60.46	−28.70	6.00	649.88
Iceland	2.18	4.80	7.50	21.17
India	1,275.61	87.90	1.19	388.05
Indonesia	295.60	97.70	1.36	155.20
Iran, Islamic Rep. of	382.09	74.80	5.60	231.82
Iraq	73.01	50.20	2.67	166.56
Ireland	44.45	39.80	11.10	632.53
Israel	68.43	106.20	10.57	3,089.95
Italy	487.28	13.20	8.40	1,617.16
Jamaica	10.74	34.70	4.09	976.89
Japan	1,259.43	12.20	9.90	3,332.95
Jordan	17.12	67.80	3.16	191.59
Kazakhstan	159.49	...	10.74	58.53
Kenya	8.79	50.70	0.27	15.15
Kiribati	0.03	40.90	0.32	42.70
Korea, Dem. People's Rep.	77.60	−68.30	3.48	643.79
Korea, Rep. of	456.75	89.10	9.62	4,588.71
Kuwait	78.60	73.40	31.13	4,411.38
Kyrgyzstan	5.33	...	1.04	26.65
Lao People's Dem. Rep.	1.25	445.20	0.22	5.30
Latvia	7.43	−60.20	3.20	115.02
Lebanon	19.00	108.60	5.42	1,826.73
Lesotho*	0.64	...	0.38	20.95
Liberia	0.46	−0.60	0.14	4.17
Libya	50.27	32.90	8.93	28.57
Liechtenstein	0.24	4.30	7.10	1500.00
Lithuania	12.29	−68.40	3.60	188.21
Luxembourg	10.69	−16.20	23.60	4133.80
Macao, China (SAR)	1.87	81.50	4.11	71,846.16
Macedonia, FYR	10.55	...	5.20	410.10
Madagascar	2.35	148.70	0.13	3.99
Malawi	0.88	47.30	0.07	7.47
Malaysia	156.68	183.00	6.41	475.01
Maldives	0.44	187.00	1.41	1,483.22
Mali	0.55	31.00	0.04	0.45
Malta	2.47	10.40	6.20	7,806.96
Martinique	1.34	−35.00	3.42	1,216.88
Mauritania	2.50	−5.20	0.87	2.44
Mauritius	3.15	115.00	2.58	1,544.12

Continued

	CO_2 emissions	% change since 1990	CO_2 emissions per caput	CO_2 emissions per km^2
	m. tons	*%*	*ton / person*	*ton / km^2*
Mexico	416.70	10.90	3.99	212.80
Micronesia, Federated States of*	0.24	...	2.20	336.14
Monaco	0.13	44.40	3.80	65,000.00
Mongolia	7.99	−20.10	3.09	5.11
Montenegro	50.02	...	4.76	489.59
Montserrat	0.06	79.40	15.99	598.04
Morocco	37.97	61.40	1.24	85.03
Mozambique	1.57	57.30	0.08	1.96
Myanmar	9.47	121.60	0.19	13.99
Namibia	2.33	38,750.00	1.17	2.83
Nauru	0.14	6.80	10.76	6,714.29
Nepal	2.95	367.60	0.11	20.08
Netherlands	176.86	11.90	11.00	4,258.81
Netherlands Antilles	4.06	236.80	22.70	5,073.75
New Caledonia	1.87	16.00	8.20	100.78
New Zealand	34.70	37.10	8.80	128.26
Nicaragua	3.92	47.90	0.74	30.13
Niger	1.21	15.00	0.09	0.95
Nigeria	52.28	15.10	0.42	56.59
Niue	0.00	0.00	2.01	11.54
Norway	43.22	25.60	9.40	112.21
Oman	32.31	214.10	12.87	104.39
Pakistan	114.36	67.80	0.75	143.65
Palau	0.24	3.80	12.30	529.41
Panama	6.03	92.50	1.93	79.92
Papua New Guinea	2.52	3.40	0.44	5.43
Paraguay	4.14	83.00	0.70	10.19
Peru	26.20	24.40	0.96	20.38
Philippines	77.10	75.30	0.96	256.98
Poland†	308.28	−35.30	...	985.91
Portugal	64.29	47.40	6.20	698.94
Puerto Rico	2.11	−82.10	0.54	237.18
Qatar	46.26	279.10	63.09	4,205.64
Republic of Moldova	7.24	...	1.71	213.88
Réunion	2.48	101.70	3.26	987.65
Romania	111.39	−39.50	5.10	467.26
Russian Federation‡	1509.00	−36.10	...	88.25
Rwanda	0.60	13.60	0.07	22.86
Saint Helena	0.01	100.00	2.46	38.96
Saint Christopher and Nevis	0.13	93.80	3.02	482.76
Saint Lucia	0.33	100.00	2.06	604.82

Continued

	CO_2 emissions	% change since 1990	CO_2 emissions per caput	CO_2 emissions per km^2
	m. tons	*%*	*ton / person*	*ton / km^2*
Saint Pierre and Miquelon	0.06	−29.30	11.32	268.60
St Vincent and the Grenadines	0.19	142.50	1.65	500.00
Samoa	0.15	19.80	0.83	53.34
São Tomé and Príncipe	0.09	35.30	0.62	95.44
Saudi Arabia	302.88	53.20	12.98	140.90
Senegal	4.85	54.50	0.44	24.63
Serbia	50.02	...	4.76	489.59
Seychelles	0.55	379.80	6.91	1,202.20
Sierra Leone	0.65	94.90	0.13	9.10
Singapore	47.88	6.10	11.35	70,109.81
Slovakia	43.05	−27.60	8.00	877.98
Slovenia	16.10	0.60	8.20	794.83
Solomon Islands	0.18	9.20	0.39	6.16
Somalia§	0.00	...	0.00	0.00
South Africa	364.85	27.60	7.78	298.81
Spain	331.76	45.30	7.90	655.66
Sri Lanka	10.32	174.10	0.51	157.31
Sudan	9.01	67.00	0.26	3.59
Suriname	2.24	23.70	5.05	13.69
Swaziland	0.96	125.20	0.92	55.11
Sweden	56.00	−0.50	6.20	124.45
Switzerland	44.72	0.80	6.20	1,083.23
Syrian Arabian Republic	49.04	36.60	2.70	264.80
Tajikistan	4.66	...	0.73	32.58
Tanzania	3.81	62.90	0.10	4.03
Thailand	246.37	156.90	3.90	480.15
Timor-Leste	0.16	...	0.20	10.96
Togo	2.20	192.60	0.38	38.74
Tonga	0.11	48.10	1.12	152.61
Trinidad and Tobago	28.70	69.40	22.12	5,594.35
Tunisia	20.91	57.40	2.11	127.80
Turkey	220.41	50.50	3.09	281.29
Turkmenistan	43.41	...	9.24	88.94
Turks and Caicos Islands‖	0.00	...	0.00	0.00
Tuvalu*	0.00	...	0.48	178.85
Uganda	1.71	110.20	0.06	7.11
Ukraine	313.14	−57.60	6.60	518.70
United Arab Emirates	135.29	147.00	33.56	1,618.24
United Kingdom	557.46	−5.30	9.40	2,295.02
USA	5,841.50	16.60	20.00	606.65
United States Virgin Islands	13.55	60.10	121.30	39,043.23

Continued

	CO$_2$ emissions	% change since 1990	CO$_2$ emissions per caput	CO$_2$ emissions per km^2
	m. tons	*%*	*ton / person*	*ton / km^2*
Uruguay	4.38	11.90	1.28	25.03
Uzbekistan	123.84	...	4.79	276.80
Vanuatu	0.09	30.90	0.44	7.30
Venezuela	144.23	22.70	5.59	158.13
Viet Nam	76.24	255.90	0.93	229.86
Western Sahara	0.24	21.80	0.75	0.90
Yemen	17.08	...	0.87	32.35
Zambia	2.20	−10.20	0.19	2.92
Zimbabwe	11.49	−31.10	0.89	29.40

Sources: UNSD Millennium Development Goals Indicators database (see http://mdgs.un.org/unsd/mdg/Data.aspx). UN Population Division. UNSD Demographic Yearbook (see: http://unstats.un.org/unsd/demographic/products/dyb/dyb2004.htm). See http://unstats.un.org/unsd/environment/air_co2_emissions.htm.

* Data refer to 1994.

† Data refer to 2002.

‡ Data refer to 1999.

§ Data refer to 1997.

|| Data refer to 1990.

1. Including part of the Neutral Zone.

2. Data refer to Serbia and Montenego.

Definitions and Technical notes: CO$_2$ emissions from energy industry, from transport, from fuel combustion in industry, services, households, etc., and industrial processes, such as the production of cement.

Changes in how land is used can also result in the emission of CO$_2$, or in the removal of CO$_2$ from the atmosphere. However, as there is not yet an agreed method for estimating this, it is not included in the figures for CO$_2$ emissions.

Burning of biomass such as wood and straw also emits CO$_2$; however, unless there has been a change in land use, it is considered that CO$_2$ emitted from biomass is removed from the air by new growth, and therefore it should not included in the total for CO$_2$.

Table 5: Freshwater resources—water withdrawals*

	Total (million cubic metres, 2000)	Per caput (cubic metres per person, 2000)
Asia (excluding Middle East)	2,147,506	631
Europe	400,266	581
Middle East and North Africa	324,646	807
Sub-Saharan Africa	113,361	173
North America	525,267	1,663
Central America and the Caribbean	100,657	603
South America	164,429	474
Oceania	26,187	900
Developed countries	1,221,192	956
Developing countries	2,583,916	545
World	3,802,320	633

* The gross amount of water extracted from any source, either permanently or temporarily, for a given use, including consumptive use, conveyance losses and return flow. Per caput withdrawals were calculated using national population data from the United Nations Population Division, 2000.
Source: World Resources Institute. 2006. *EarthTrends: The Environmental Information Portal.* Available at http://earthtrends.wri.org. Washington, DC: World Resources Institute. Data sources: Food and Agriculture Organization of the United Nations (FAO—AQUASTAT), Population Division of the Department of Economic and Social Affairs of the United Nations Secretariat.

Table 6: Water resources and freshwater ecosystems—water withdrawals: withdrawals as a percentage of internal water resources

Country	2000
Afghanistan	42.3
Albania	6.4
Algeria	54.0
Angola	0.2
Antigua and Barbuda[1]	9.6
Argentina	10.6
Armenia	32.5
Australia	4.9
Austria	3.8
Azerbaijan	212.6
Bahrain	7,500.0
Bangladesh	75.6
Barbados	112.5
Belarus	7.5
Belgium[2]	75.3
Belize	0.9
Benin	1.3
Bhutan	0.4
Bolivia	0.5
Botswana	8.1
Brazil	1.1
Brunei Darussalam[3]	1.1
Bulgaria	50.0
Burkina Faso	6.4
Burundi	2.9
Cambodia	3.4
Cameroon	0.4
Canada	1.6
Cape Verde	7.3
Côte d'Ivoire	1.2
Central African Republic	–
Chad	1.5
Chile	1.4
China	22.4
Colombia	0.5
Comoros	0.8
Congo, Republic of the	–
Congo, Dem. Rep. of the	–
Costa Rica	2.4
Cuba	21.5

Continued

Country	2000
Cyprus	30.8
Czech Republic	19.6
Denmark	21.2
Djibouti	6.3
Dominican Republic	16.1
Ecuador	3.9
Egypt	3,794.4
El Salvador	7.2
Equatorial Guinea	0.4
Eritrea[4]	20.8
Estonia	1.2
Ethiopia	4.6
Fiji	0.2
Finland	2.3
France	22.4
Gabon	0.1
Gambia	1.0
Georgia	6.2
Germany	44.0
Ghana	3.2
Greece	13.4
Guatemala	1.8
Guinea	0.7
Guinea-Bissau	1.1
Guyana	0.7
Haiti	7.6
Honduras	0.9
Hungary	127.3
Iceland	0.1
India	51.2
Indonesia	2.9
Iran, Islamic Rep. of	56.7
Iraq	121.3
Ireland	2.3
Israel	273.3
Italy	24.3
Jamaica	4.4
Japan	20.6
Jordan	148.5
Kazakhstan	46.4
Kenya	7.6
Korea, Dem. People's Rep.	13.5

Continued

Country	2000
Korea, Rep. of	28.7
Kyrgyzstan	21.7
Lao People's Dem. Rep.	1.6
Latvia	1.8
Lebanon	28.8
Lesotho	1.0
Liberia	0.1
Libyan Arab Jamahiriya	711.3
Lithuania	1.7
Luxembourg[5]	4.0
Madagascar	4.4
Malawi	6.3
Malaysia	1.6
Maldives[6]	11.2
Mali	10.9
Malta	100.0
Mauritania	425.0
Mauritius[7]	26.4
Mexico	19.1
Moldova, Republic of	231.0
Mongolia	1.3
Morocco	43.4
Mozambique	0.6
Myanmar	3.8
Namibia	4.9
Nepal	5.1
Netherlands	72.2
New Zealand	0.6
Nicaragua	0.7
Niger	62.3
Nigeria	3.6
Norway	0.6
Oman	138.1
Pakistan	323.3
Panama	0.6
Papua New Guinea	–
Paraguay	0.5
Peru	1.2
Philippines	6.0
Poland	30.2
Portugal	29.6
Qatar	568.6

Continued

Country	2000
Romania	54.8
Russian Federation	1.8
Rwanda	1.6
São Tomé & Príncipe[8]	0.3
Saudi Arabia	721.7
Senegal	8.6
Sierra Leone	0.2
Singapore[9]	31.7
Somalia[10]	55.0
South Africa	27.9
Spain	32.0
Sri Lanka	25.2
Sudan	124.4
Suriname	0.8
Swaziland	39.5
Sweden	1.7
Switzerland	6.4
Syrian Arab Republic	285.0
Tajikistan	18.0
Tanzania	6.2
Thailand	41.5
Togo	1.5
Trinidad and Tobago	8.1
Tunisia	62.9
Turkey	16.5
Turkmenistan	1,812.5
Uganda	0.8
Ukraine	70.7
United Arab Emirates	1,533.3
United Kingdom	6.6
USA	17.1
Uruguay	5.3
Uzbekistan	357.0
Venezuela	1.2
Viet Nam	19.5
Yemen	161.7
Zambia	2.2
Zimbabwe	34.3
World	8.9

Notes:
1. Antigua and Barbuda: 1990 value.

2. Belgium: 1980 value.
3. Brunei Darussalam: 1995 value.
4. Eritrea: 2005 value.
5. Luxembourg: 1985 value.
6. Maldives: 1985 value.
7. Mauritius: 2005 value.
8. São Tomé & Príncipe: 1995 value.
9. Singapore: 1975 value.
10. Somalia: 2005 value.

Source: World Resources Institute. 2007. EarthTrends: Environmental Information. Available at http://earthtrends.wri.org. Washington, DC: World Resources Institute.

Food and Agriculture Organization of the United Nations (FAO) Land and Water Development Division. 2005. AQUASTAT Information System on Water and Agriculture: Online database. Rome: FAO. Available on-line at http://www.fao.org/waicent/faoinfo/agricult/agl/aglw/aquastat/dbase/index.stm.

Technical Notes: Withdrawals as a percent of internal water resources are the proportion of internal renewable water resources withdrawn on an annual basis.

Annual total water withdrawals is the gross amount of water extracted from any source, either permanently or temporarily, for a given use. It can be either diverted towards distribution networks or directly used. It includes consumptive use, conveyance losses, and return flow.

Internal Renewable Water Resources (IRWR) is comprised of the average annual flow of rivers and recharge of groundwater (aquifers) generated from endogenous (internal) precipitation. Natural incoming flows originating outside a country's borders are not included in the total. Even though IRWR measures a combination of surface and groundwater resources, it is typically less than the sum of the two because of overlap–water resources that are counted with both surface and groundwater.

IRWR is calculated as follows:

IRWR = surface water resources + groundwater resources – overlap.

Methodology: Withdrawals as a percent of internal water resources is calculated by dividing total water withdrawals by total internal renewable water resources.

Table 7: Water resources and freshwater ecosystems—actual renewable water resources per caput

Cubic metres (m^3) per person per year

Country	2006
Afghanistan	2,091.20
Albania	13,250.70
Algeria	349.80
Andorra	–
Angola	9,024.40
Antigua and Barbuda	634.10
Argentina	20,800.30
Armenia	3,501.50
Australia[1]	24,157.90
Austria	9,469.80
Azerbaijan	3,574.00
Bahamas	61.20
Bahrain	157.00
Bangladesh	8,381.80
Barbados	296.30
Belarus	5,979.40
Belgium	1,753.40
Belize	67,472.70
Benin	3,032.60
Bhutan	42,967.00
Bolivia	66,552.40
Bosnia and Herzegovina	9,585.90
Botswana	6,954.50
Brazil	43,587.80
Brunei Darussalam	22,251.30
Bulgaria	2,776.70
Burkina Faso	916.80
Burundi	1,600.20
Cambodia	33,176.10
Cameroon	17,197.80
Canada	89,111.30
Cape Verde	578.00
Côte d'Ivoire	4,396.90
Central African Republic	35,279.70
Chad	4,286.30
Chile	55,997.60
China[2]	2,137.30
Colombia	46,068.40
Comoros[3]	1,465.20

Continued

Cubic metres (m³) per person per year

Country	2006
Congo, Rep. of the	221,034.70
Congo, Dem. Rep. of the	21,628.50
Costa Rica	25,551.30
Croatia	23,156.30
Cuba	3,375.20
Cyprus	923.10
Czech Republic	1,288.10
Denmark	1,101.70
Djibouti	371.70
Dominican Republic	2,327.30
Ecuador	31,626.80
Egypt	772.80
El Salvador	3,604.80
Equatorial Guinea	50,485.40
Eritrea	1,381.60
Estonia	9,666.40
Ethiopia	1,538.70
Fiji	33,430.90
Finland[4]	20,904.60
France	3,354.60
French Guiana	701,570.70
Gabon	116,643.00
Gambia	5,141.40
Gaza Strip	14.70
Georgia	14,282.80
Germany	1,861.80
Ghana	2,358.60
Greece	6,665.20
Greenland	10,578,947.40
Guatemala	8,618.20
Guinea	23,534.30
Guinea-Bissau	18,971.80
Guyana	320,478.70
Haiti	1,621.40
Honduras	13,030.30
Hungary	10,326.70
Iceland	572,390.60
India	1,694.10
Indonesia	12,587.30
Iran, Islamic Rep. of	1,955.40

Continued

Cubic metres (m^3) per person per year

Country	2006
Iraq	2,552.20
Ireland	12,351.50
Israel	243.90
Italy	3,290.30
Jamaica	3,532.70
Japan	3,353.60
Jordan	150.80
Kazakhstan	7,400.10
Kenya	874.50
Korea, Dem. People's Rep. of	3,415.60
Korea, Rep.	1,452.60
Kuwait	7.20
Kyrgyzstan	3,864.80
Lao People's Dem. Rep.	55,059.40
Latvia	15,446.20
Lebanon	1,219.40
Lesotho	1,687.30
Liberia	69,129.90
Libyan Arab Jamahiriya	100.50
Lithuania	7,287.10
Luxembourg	6,581.70
Macedonia, FYR	3,141.90
Madagascar	17,639.40
Malawi	1,312.50
Malaysia	22,484.10
Maldives	89.00
Mali	7,184.90
Malta	125.30
Mauritania	3,609.90
Mauritius[5]	2,190.30
Mexico	4,220.80
Moldova, Republic	2,777.10
Mongolia	12,989.90
Morocco	907.90
Mozambique	10,770.40
Myanmar	20,498.40
Namibia	8,633.00
Nepal	7,594.50
Netherlands	5,560.00
New Zealand	80,482.40

Continued

Cubic metres (m^3) per person per year

Country	2006
Nicaragua	35,123.20
Niger	2,332.60
Nigeria	2,129.90
Norway[6]	82,274.40
Oman	377.10
Pakistan	1,381.30
Panama	45,006.10
Papua New Guinea	133,477.80
Paraguay	53,324.90
Peru	67,406.60
Philippines	5,670.20
Poland	1,600.00
Portugal	6,514.90
Puerto Rico	1,785.30
Qatar	63.20
Réunion	6,281.40
Romania	9,798.40
Russian Federation	31,621.60
Rwanda	1,029.30
Saint Christopher and Nevis	558.10
São Tomé & Príncipe	13,625.00
Saudi Arabia	95.30
Senegal	3,250.70
Serbia and Montenegro	19,862.80
Sierra Leone	28,174.00
Singapore	137.00
Slovakia	9,276.10
Slovenia	16,210.60
Solomon Islands	91,224.50
Somalia	1,730.20
South Africa	1,050.60
Spain	2,570.40
Sri Lanka	2,391.00
Sudan	1,743.60
Suriname	269,911.50
Swaziland	4,382.90
Sweden	19,184.10
Switzerland	7,365.10
Syrian Arab Republic	1,345.80
Tajikistan	2,424.50

Continued

Cubic metres (m^3) per person per year

Country	2006
Tanzania	2,466.90
Thailand	6,330.00
Togo	2,331.10
Trinidad and Tobago	2,933.50
Tunisia	450.00
Turkey	2,879.00
Turkmenistan	5,045.90
Uganda	2,210.50
Ukraine	3,034.60
United Arab Emirates	32.20
United Kingdom	2,456.30
USA	10,135.20
Uruguay	39,862.30
Uzbekistan	1,868.40
Venezuela	45,310.50
Viet Nam	10,442.60
Yemen	189.50
Zambia	8,869.40
Zimbabwe	1,528.50
World[7]	8,462.00

Notes:

1. Australia: population data includes Christmas Island, Cocos (Keeling) Islands, and Norfolk Island.
2. China: population data do not include Hong Kong and Macao, Special Administrative Regions (SAR) of China.
3. Comoros: population data includes the island of Mayotte.
4. Finland: population data include Åland Islands.
5. Mauritius: population data include Agalega, Rodrigues, and Saint Brandon.
6. Norway: population data include Svalbard and Jan Mayen Island.
7. World: value calculated by World Resources Institute.

Source: World Resources Institute. 2007. EarthTrends: Environmental Information. Available at http://earthtrends.wri.org. Washington, DC: World Resources Institute (WRI).

Food and Agriculture Organization of the United Nations (FAO) Land and Water Development Division. 2005. AQUASTAT Information System on Water and Agriculture: Online database. Rome: FAO. Available on-line at http://www.fao.org/waicent/faoinfo/agricult/agl/aglw/aquastat/dbase/index.stm.

Population Division of the Department of Economic and Social Affairs of the United Nations Secretariat, 2005. World Population Prospects: The 2004 Revision. Dataset on CD-ROM. New York: United Nations. Available on-line at http://www.un.org/esa/population/ordering.htm.

Technical Notes: Per Caput Actual Renewable Water Resources gives the maximum theoretical amount of water actually available, on a per person basis, for each country. In reality, a portion of this water may be inaccessible to humans. Actual renewable water resources are defined as the sum of internal renewable resources (IRWR) and external renewable resources (ERWR), taking into consideration the quantity of flow reserved to upstream and downstream countries through formal or informal agreements or treaties and possible reduction of external flow due to upstream water abstraction.

Internal renewable water resources (IRWR) are comprised of the average annual flow of rivers and recharge of groundwater (aquifers) generated from endogenous (internal) precipitation. Even though IRWR measures a combination of surface and groundwater resources, it is typically less than the sum of the two because of overlap–water resources that are common to both surface and groundwater.

ERWR are the portion of the country's renewable water resources which is not generated within the country. The ERWR include inflows from upstream countries (groundwater and surface water), and part of the water of border lakes or rivers. Per caput water resources are calculated by WRI using 2006 population estimates from the United Nations Population Division.

Data are labeled as being from the year 2006, since the per capita actual renewable water resources were calculated with population estimates from 2006. Original data were actually collected over a period of 15 to 25 years and compiled in 2003 by the AQUASTAT global information system of water and agriculture.

Methodology: Per capita actual water resources were calculated by WRI using the United Nations Population Division's World Population Prospects: The 2004 Revision. The computation of actual renewable water resources requires the assessment of both internal and external water resources. Internal renewable water resources are computed by adding up average annual surface runoff and groundwater recharge occurring within a country's borders. External flows include both natural and actual incoming flows.

While AQUASTAT represents the most complete and careful compilation of water resources statistics to date, freshwater data are generally of poor quality. Information sources are various but rarely complete. Access to information on water resources is still sometimes restricted for reasons related to sensitivity at the regional level. The accuracy and reliability of the information vary greatly among regions, countries, and categories of information, as does the year in which the information was gathered. As a result, no consistency can be ensured at the regional level on the duration and dates of the period of reference.

Table 8: Threatened species in each country (totals by taxonomic group) (Red List 2007)

(Threatened species are those listed as **Critically Endangered (CR)**, **Endangered (EN) or Vulnerable (VU)**.)

	Mammals	Birds	Reptiles	Amphibians	Fishes	Molluscs	Other Invertebrates	Plants	Total
AFRICA									
North Africa									
Algeria	15	10	7	3	22	0	14	3	74
Egypt	14	10	11	0	23	0	1	2	61
Libya	9	4	5	0	13	0	0	1	32
Morocco	17	10	9	2	29	0	9	2	78
Tunisia	14	7	4	1	19	0	7	0	52
Western Sahara	9	1	0	0	19	0	1	0	30
Sub-Saharan Africa									
Angola	13	18	4	0	22	4	1	26	88
Benin	11	4	4	0	15	0	0	14	48
Botswana	8	8	0	0	2	0	0	0	18
Burkina Faso	8	4	1	0	0	0	0	2	15
Burundi	11	8	0	6	18	1	4	2	50
Cameroon	42	15	2	53	43	1	1	355	512
Cape Verde	3	4	0	0	18	0	0	2	27
Central African Republic	11	5	1	0	0	0	0	15	32
Chad	13	6	1	0	0	1	0	2	23
Comoros	3	9	2	0	5	0	4	5	28
Congo, Rep. of the	14	3	1	0	15	1	3	35	72
Congo, Dem. Rep.	28	31	3	13	25	13	13	65	191
Côte d'Ivoire	25	12	3	13	19	1	0	105	178
Djibouti	7	7	0	0	14	0	0	2	30
Equatorial Guinea	17	5	3	4	13	0	0	63	105
Eritrea	12	7	6	0	13	0	0	3	41

STATISTICS

Continued

	Mammals	Birds	Reptiles	Amphibians	Fishes	Molluscs	Other Invertebrates	Plants	Total
Ethiopia	38	21	1	9	2	3	12	22	108
Gabon	12	5	2	3	21	0	0	108	151
Gambia	9	5	1	0	16	0	0	4	35
Ghana	18	8	3	10	17	0	0	117	173
Guinea	21	12	1	5	19	0	3	22	83
Guinea-Bissau	9	1	1	0	18	0	0	4	33
Kenya	32	27	5	6	70	16	16	103	275
Lesotho	3	5	0	0	1	0	2	1	12
Liberia	21	11	3	4	19	1	1	46	106
Madagascar	47	35	20	55	73	24	8	280	542
Malawi	7	12	0	5	101	9	7	14	155
Mali	13	6	1	0	1	0	0	6	27
Mauritania	11	8	2	0	22	0	1	0	44
Mauritius	4	11	7	0	11	27	5	88	153
Mayotte	1	4	2	0	1	0	1	0	9
Mozambique	14	21	5	3	45	4	1	46	139
Namibia	10	21	3	1	20	0	0	24	79
Niger	12	5	0	0	2	0	1	2	22
Nigeria	29	12	3	13	21	0	1	171	250
Rwanda	17	10	0	8	9	0	5	3	52
Réunion	4	6	3	0	6	14	2	16	51
Saint Helena	1	18	1	0	11	0	2	26	59
São Tomé and Príncipe	3	10	1	3	9	1	1	35	63
Senegal	13	8	6	0	28	0	0	7	62
Seychelles	3	10	10	6	14	2	3	45	93
Sierra Leone	15	10	3	2	16	0	2	47	95
Somalia	15	12	2	0	25	1	0	17	72

Continued

	Mammals	Birds	Reptiles	Amphibians	Fishes	Molluscs	Other Invertebrates	Plants	Total
South Africa	29	36	19	21	66	24	128	73	396
Sudan	17	13	2	0	13	0	2	17	64
Swaziland	6	7	0	0	3	0	0	11	27
Tanzania	34	39	5	41	137	17	26	240	539
Togo	10	2	2	3	16	0	0	10	43
Uganda	28	17	0	6	54	10	16	38	169
Zambia	12	10	0	1	10	3	2	8	46
Zimbabwe	10	12	0	6	3	0	4	17	52
ANTARCTIC									
Antarctica	0	4	0	0	0	0	0	0	4
Bouvet Island	0	1	0	0	1	0	0	0	2
French Southern Territories	2	13	0	0	2	0	0	0	17
Heard Island and McDonald Islands	0	11	0	0	1	0	0	0	12
South Georgia and the South Sandwich Islands	1	7	0	0	0	0	0	0	8
ASIA									
East Asia									
China	83	86	31	85	60	1	5	446	797
Hong Kong, China SAR	1	15	1	3	14	1	2	6	43
Japan	37	39	11	20	40	25	18	12	202
Korea, Dem. People's Rep.	11	19	0	1	11	0	2	3	47
Korea, Rep. of	10	27	0	1	14	0	2	0	54
Macao	0	3	0	0	6	0	0	0	9
Mongolia	14	20	0	0	1	0	3	0	38
Taiwan, Province of China	13	20	8	9	37	1	0	78	166

269

Continued

	Mammals	Birds	Reptiles	Amphibians	Fishes	Molluscs	Other Invertebrates	Plants	Total
North Asia									
Belarus	6	3	0	0	0	0	8	0	17
Moldova, Republic of	5	9	1	0	9	0	4	0	28
Russian Federation	45	51	6	0	22	1	28	7	160
Ukraine	16	12	2	0	14	0	14	1	59
South and South-East Asia									
Bangladesh	29	26	21	1	12	0	0	12	101
Bhutan	22	16	1	1	0	0	1	7	48
British Indian Ocean Territory	0	0	2	0	7	0	0	1	10
Brunei Darussalam	15	21	4	3	7	0	0	99	149
Cambodia	27	24	11	3	17	0	0	31	113
Disputed Territory	0	0	0	0	1	0	0	0	1
India	89	75	25	63	39	2	20	247	560
Indonesia	146	116	27	33	111	3	28	386	850
Lao People's Dem. Rep.	34	22	11	4	6	0	0	21	98
Malaysia	50	40	21	46	47	19	2	686	911
Maldives	1	0	2	0	11	0	0	0	14
Myanmar	39	39	22	0	16	1	1	38	156
Nepal	32	31	6	3	0	0	0	7	79
Philippines	51	67	9	48	58	3	17	213	466
Singapore	4	13	4	0	22	0	1	54	98
Sri Lanka	21	13	8	52	31	0	52	280	457
Thailand	38	43	22	3	50	1	0	86	243
Timor-Leste	1	5	1	0	4	0	0	0	11
Viet Nam	43	38	25	15	31	0	0	146	298
West and Central Asia									
Afghanistan	16	14	1	1	0	0	1	2	35

Continued

	Mammals	Birds	Reptiles	Amphibians	Fishes	Molluscs	Other Invertebrates	Plants	Total
Armenia	11	12	5	0	1	0	6	1	36
Azerbaijan	11	13	5	0	5	0	4	0	38
Bahrain	2	4	4	0	6	0	0	0	16
Cyprus	4	4	4	0	11	0	0	7	30
Georgia	13	8	7	1	8	0	9	0	46
Iran, Islamic Rep. of	24	18	8	4	16	0	5	1	76
Iraq	11	18	2	1	6	0	2	0	40
Israel	15	12	10	0	30	5	7	0	79
Jordan	12	8	5	0	14	0	4	0	43
Kazakhstan	19	20	2	1	9	0	4	16	71
Kuwait	4	8	1	0	10	0	0	0	23
Kyrgyzstan	9	8	2	0	0	0	3	14	36
Lebanon	9	6	6	0	14	0	3	0	38
Oman	12	9	4	0	21	0	4	6	56
Pakistan	23	26	9	0	20	0	0	2	80
Palestinian Territory, Occupied	0	4	4	0	0	0	1	0	9
Qatar	1	4	1	0	7	0	0	0	13
Saudi Arabia	12	14	2	0	15	0	2	3	48
Syrian Arab Republic	10	11	6	0	26	0	6	0	59
Tajikistan	10	9	1	0	5	0	2	14	41
Turkey	18	15	13	9	54	0	12	3	124
Turkmenistan	15	15	1	0	8	0	5	3	47
United Arab Emirates	7	8	1	0	9	0	2	0	27
Uzbekistan	10	15	2	0	5	0	1	15	48
Yemen	8	13	2	1	17	2	4	159	206
EUROPE									
Albania	2	6	4	2	27	0	4	0	45

271

Continued

	Mammals	Birds	Reptiles	Amphibians	Fishes	Molluscs	Other Invertebrates	Plants	Total
Andorra	1	0	1	0	1	1	3	0	7
Austria	6	5	1	0	7	22	21	4	66
Belgium	8	1	0	0	8	4	8	1	30
Bosnia and Herzegovina	8	6	2	1	28	0	10	1	56
Bulgaria	13	12	2	0	13	0	7	0	47
Croatia	7	11	2	2	42	0	14	1	79
Czech Republic	7	5	0	0	9	2	16	4	43
Denmark	3	3	0	0	11	1	10	3	31
Estonia	5	3	0	0	2	0	4	0	14
Faroe Islands	3	0	0	0	8	0	0	0	11
Finland	4	3	0	0	2	1	9	1	20
France	15	5	5	2	27	34	29	7	124
Germany	9	4	0	0	16	9	21	12	71
Gibraltar	1	3	0	0	9	2	0	0	15
Greece	11	10	5	5	50	1	13	11	106
Greenland	8	0	0	0	6	0	0	1	15
Guernsey	0	0	0	0	2	0	0	0	2
Hungary	9	9	1	0	10	1	25	1	56
Iceland	6	0	0	0	11	0	0	0	17
Ireland	3	1	0	0	8	1	2	1	16
Isle of Man	0	0	0	0	2	0	0	0	2
Italy	12	7	5	6	31	16	42	19	138
Jersey	0	0	0	0	2	0	0	0	2
Latvia	5	4	0	0	4	1	9	0	23
Liechtenstein	2	0	0	0	0	0	4	0	6
Lithuania	6	4	0	0	4	0	6	0	20
Luxembourg	3	0	0	0	0	2	2	0	7

Continued

	Mammals	Birds	Reptiles	Amphibians	Fishes	Molluscs	Other Invertebrates	Plants	Total
Macedonia, FYR	9	10	2	0	8	0	5	0	34
Malta	1	3	0	0	13	3	0	3	23
Monaco	0	0	0	0	10	0	0	0	10
Montenegro	6	9	2	1	20	0	11	0	49
Netherlands	10	1	0	0	9	1	5	0	26
Norway	10	2	0	0	11	1	8	2	34
Poland	13	5	0	0	4	1	15	4	42
Portugal	15	8	2	0	39	67	16	16	163
Romania	15	12	2	0	13	0	22	1	65
Serbia	8	10	0	0	8	0	16	1	43
Slovakia	8	7	1	0	9	6	13	2	46
Slovenia	7	3	1	2	25	0	42	0	80
Spain	20	15	17	5	51	27	35	49	219
Svalbard and Jan Mayen	6	0	0	0	2	0	0	0	8
Sweden	5	3	0	0	9	1	12	3	33
Switzerland	4	2	0	1	8	0	29	3	47
United Kingdom	9	3	0	0	16	2	8	13	51
NORTH AND CENTRAL AMERICA									
Caribbean Islands									
Anguilla	1	0	4	0	15	0	0	3	23
Antigua and Barbuda	1	1	5	0	15	0	0	4	26
Aruba	2	1	3	0	16	0	1	0	23
Bahamas	5	5	6	0	20	0	1	5	42
Barbados	1	1	4	0	14	0	0	2	22
Bermuda	2	1	2	0	13	0	25	4	47
Cayman Islands	0	1	5	0	14	1	0	2	23
Cuba	11	17	7	47	28	0	5	163	278

Continued

	Mammals	Birds	Reptiles	Amphibians	Fishes	Molluscs	Other Invertebrates	Plants	Total
Dominica	2	3	4	2	16	0	0	11	38
Dominican Republic	5	14	10	31	15	0	6	30	111
Grenada	1	1	4	1	16	0	0	3	26
Guadeloupe	6	2	5	3	15	1	0	8	40
Haiti	4	13	9	46	16	0	3	29	120
Jamaica	5	10	8	17	16	1	5	209	270
Martinique	1	3	5	2	15	0	0	9	36
Montserrat	2	2	3	1	15	0	0	4	27
Netherlands Antilles	2	1	6	0	16	0	0	2	27
Puerto Rico	3	8	8	13	14	0	1	53	100
Saint Christopher and Nevis	1	1	5	1	15	0	0	2	25
Saint Lucia	2	5	6	0	14	0	0	6	33
Saint Vincent and the Grenadines	3	2	4	1	15	0	0	5	30
Trinidad and Tobago	1	2	5	9	21	0	0	1	39
Turks and Caicos Islands	1	2	5	0	13	0	0	2	23
Virgin Islands, British	0	1	6	2	14	0	0	10	33
Virgin Islands, US	1	1	5	2	12	0	0	11	32
Central America									
Belize	5	3	5	6	24	0	1	30	74
Costa Rica	12	17	8	62	20	0	12	111	242
El Salvador	3	4	6	9	7	0	0	26	55
Guatemala	7	11	14	76	18	2	5	84	217
Honduras	9	7	11	55	19	0	1	110	212
Mexico	72	59	95	198	115	5	35	261	840
Nicaragua	5	9	8	10	22	2	3	39	98
Panama	17	19	7	55	21	0	2	194	315

Continued

	Mammals	Birds	Reptiles	Amphibians	Fishes	Molluscs	Other Invertebrates	Plants	Total
North America									
Canada	17	18	3	1	26	2	10	1	78
Saint Pierre and Miquelon	0	1	0	0	1	0	0	0	2
USA	41	74	32	53	166	273	298	242	1,179
SOUTH AMERICA									
South America									
Argentina	29	49	5	29	30	0	10	42	194
Bolivia	25	31	2	21	0	0	1	71	151
Brazil*	73	122	22	25	66	21	14	382	725
Chile	22	33	0	20	18	0	2	39	134
Colombia	38	87	15	209	31	0	2	222	604
Ecuador	33	68	10	163	15	48	3	1,838	2,178
Falkland Islands (Malvinas)	4	10	0	0	5	0	0	5	24
French Guiana	9	0	7	3	22	0	0	16	57
Guyana	11	3	6	6	23	0	1	22	72
Paraguay	9	28	2	0	0	0	0	10	49
Peru	46	94	6	80	10	0	2	274	512
Suriname	11	0	6	2	21	0	0	26	66
Uruguay	6	25	3	4	27	0	1	1	67
Venezuela	24	26	13	69	31	0	3	68	234
OCEANIA									
Oceania									
American Samoa	3	9	2	0	7	5	0	1	27
Australia	64	50	38	47	87	175	107	55	623
Christmas Island	0	5	3	0	6	0	0	1	15
Cocos (Keeling) Islands	0	0	1	0	7	0	0	0	8

Continued

	Mammals	Birds	Reptiles	Amphibians	Fishes	Molluscs	Other Invertebrates	Plants	Total
Cook Islands	1	15	2	0	7	0	0	1	26
Fiji	5	10	6	1	10	3	0	66	101
French Polynesia	3	32	1	0	12	29	0	47	124
Guam	2	9	2	0	9	6	0	4	32
Kiribati	0	5	1	0	6	1	0	0	13
Marshall Islands	1	2	2	0	9	1	0	0	15
Micronesia, Federated States of	5	8	2	0	12	4	0	5	36
Nauru	0	2	0	0	5	0	0	0	7
New Caledonia	6	15	2	0	16	11	1	219	270
New Zealand	8	70	12	4	16	5	9	21	145
Niue	0	8	1	0	6	0	0	1	16
Norfolk Island	0	17	2	0	2	12	0	1	34
Northern Mariana Islands	2	13	2	0	9	4	0	5	35
Palau	3	1	2	0	11	5	0	4	26
Papua New Guinea	58	31	9	10	38	2	10	142	300

* It should be noted that for certain amphibian species endemic to Brazil, there was not time to reach agreement on the Red List Categories between the Global Amphibian Assessment (GAA) Co-ordinating Team, and the experts on the species in Brazil. The 2006 figures for amphibians displayed here are those that were agreed at the GAA Brazil workshop in April 2003. However, in the subsequent consistency check conducted by the GAA Co-ordinating Team, many of the assessments were found to be inconsistent with the approach adopted elsewhere in the world, and a 'consistent Red List Category' was also assigned to these species. There was not time to agree these 'consistent Red List Categories' with the Brazilian experts before the release of the *IUCN Red List*, therefore the original workshop assessments are retained here.

Source: IUCN 2007. 2007 IUCN Red List of Threatened Species. www.iucnredlist.org. Downloaded on 13 September 2007.

When interpreting these tables, please note that the number of threatened species per country refers to species assessed as threatened at the global level. Many countries will have Red Lists of their own at the national level, and in some cases species may be threatened at the national leval although they are of Least Concern (not threatened) at the global level.

Table 9: Numbers of threatened species by major groups of organisms (1996–2007)

	Number of described species	Number of species evaluated by 2007	Number of threatened species in 1996/98	Number of threatened species in 2000	Number of threatened species in 2002	Number of threatened species in 2003	Number of threatened species in 2004	Number of threatened species in 2006	Number of threatened species in 2007	Number threatened in 2007, as % of species described	Number threatened in 2007, as % of species evaluated**
Vertebrates											
Mammals	5,416	4,863	1,096	1,130	1,137	1,130	1,101	1,093	1,094	20%	22%
Birds	9,956	9,956	1,107	1,183	1,192	1,194	1,213	1,206	1,217	12%	12%
Reptiles	8,240	1,385	253	296	293	293	304	341	422	5%	30%
Amphibians*	6,199	5,915	124	146	157	157	1,770	1,811	1,808	29%	31%
Fishes	30,000	3,119	734	752	742	750	800	1,171	1,201	4%	39%
Subtotal	59,811	25,238	3,314	3,507	3,521	3,524	5,188	5,622	5,742	10%	23%
Invertebrates											
Insects	950,000	1,255	537	555	557	553	559	623	623	0.07%	50%
Molluscs	81,000	2,212	920	938	939	967	974	975	978	1.21%	44%
Crustaceans	40,000	553	407	408	409	409	429	459	460	1.15%	83%
Corals	2,175	13	–	–	–	–	–	–	5	0.23%	38%
Others	130,200	83	27	27	27	30	30	44	42	0.03%	51%
Subtotal	1,203,375	4,116	1,891	1,928	1,932	1,959	1,992	2,101	2,108	0.18%	51%
Mosses	15,000	92	–	80	80	80	80	80	79	0.53%	86%
Ferns and allies	13,025	211	–	–	–	111	140	139	139	1%	66%
Gymnosperms	980	909	142	141	142	304	305	306	321	33%	35%
Dicotyledons	199,350	9,622	4,929	5,099	5,202	5,768	7,025	7,086	7,121	4%	74%
Monocotyledons	59,300	1,149	257	291	290	511	771	779	778	1%	68%
Green Algae	3,715	2	–	–	–	–	–	–	0	0.00%	0%
Red Algae	5,956	58	–	–	–	–	–	–	9	0.15%	16%
Subtotal	297,326	12,043	5,328	5,611	5,714	6,774	8,321	8,390	8,447	3%	70%
Others											
Lichens	10,000	2	–	–	–	2	2	2	2	0.02%	100%
Mushrooms	16,000	1	–	–	–	–	–	1	1	0.01%	100%
Brown Algae	2,849	15	–	–	–	–	–	–	6	0.21%	40%
Subtotal	28,849	18	–	–	–	2	2	3	9	0.03%	50%
TOTAL	1,589,361	41,415	10,533	11,046	11,167	12,259	15,503	16,116	16,306	1%	39%

277

NOTES:

1) *It should be noted that for certain species endemic to Brazil, there was not time to reach agreement on the Red List Categories between the Global Amphibian Assessment (GAA) Co-ordinating Team, and the experts on the species in Brazil. The 2004–2007 figures for amphibians displayed here are those that were agreed at the GAA Brazil workshop in April 2003. However, in the subsequent consistency check conducted by the GAA Co-ordinating Team, many of the assessments were found to be inconsistent with the approach adopted elsewhere in the world, and a 'consistent Red List Category' was also assigned to these species. There was not time to agree these 'consistent Red List Categories' with the Brazilian experts before the release of the IUCN Red List; therefore the original workshop assessments are retained here. However, in order to retain comparability between results for amphibians with those for other taxonomic groups, the data used in the Global Species Assessment (Baillie et al. 2004) are based on the 'consistent Red List Categories'. Therefore, figures in Table 1 above will not completely match figures in Table 2.1 in the Global Species Assessment.

2) **Apart from the mammals, birds, amphibians and gymnosperms (i.e., those groups completely or almost completely evaluated), the figures in the last column are gross over-estimates of the percentage threatened due to biases in the assessment process towards assessing species that are thought to be threatened, species for which data are readily available, and under-reporting of Least Concern species. The true value for the percentage threatened lies somewhere in the range indicated by the two right-hand columns. In most cases this represents a very broad range. For example, the true percentage of threatened insects lies somewhere between 0.07% and 50%. Hence, although 39% of all species on the IUCN Red List are listed as threatened, this figure needs to be treated with extreme caution given the biases described above.

3) Mosses include the true mosses (Bryopsida), the hornworts (Anthocerotopsida), and liverworts (Marchantiopsida); while the ferns and allies include the club mosses (Lycopodiopsida), spike mosses (Sellaginellopsida), quillworts (Isoetopsida), and true ferns (Polypodiopsida).

4) Seaweeds are included in the green algae (Chorophyta), red algae (Rhodophyta), and brown algae (Ochrophyta).

5) Threatened species are those listed as Critically Endangered (CR), Endangered (EN) or Vulnerable (VU).

6) The numbers and percentages of species threatened in each group do not mean that the remainder are all not threatened (i.e., are Least Concern). There are a number of species in many of the groups listed as Near Threatened or Data Deficient (see Tables 3a and 3b). These figures also need to be considered in relation to the number of species evaluated as shown in column two (see note 2 above).

7) The plant figures do not include species from the 1997 IUCN Red List of Threatened Plants (Walter and Gillett 1998) as those were all assessed using the pre-1994 IUCN system of threat categorization. Hence the figures for numbers of threatened plants are very much lower when compared to the 1997 results. The results from this Red List and the 1997 Plants Red List should be combined together when reporting on threatened plants.

Source: IUCN 2007. 2007 IUCN Red List of Threatened Species. www.iucnredlist.org. Downloaded on 13 September 2007.

When interpreting these tables, please note that the number of threatened species per country refers to species assessed as threatened at the global level. Many countries will have Red Lists of their own at the national level, and in some cases species may be threatened at the national leval although they are of Least Concern (not threatened) at the global level.

Table 10: Forest extent: total forest area

Units: '000 hectares

Country	1990	2000	2005
Afghanistan	1,309	1,015	867
Albania	789	769	794
Algeria	1,790	2,144	2,277
American Samoa	18	18	18
Andorra	16	16	16
Angola	60,976	59,728	59,104
Antigua and Barbuda	9	9	9
Argentina	35,262	33,770	33,021
Armenia	346	305	283
Aruba	–	–	–
Australia	167,904	164,645	163,678
Austria	3,776	3,838	3,862
Azerbaijan	936	936	936
Bahamas	515	515	515
Bahrain	–	–	–
Bangladesh	882	884	871
Barbados	2	2	2
Belarus	7,376	7,848	7,894
Belgium	677	667	667
Belize	1,653	1,653	1,653
Benin	3,322	2,675	2,351
Bermuda	1	1	1
Bhutan	3,035	3,141	3,195
Bolivia	62,795	60,091	58,740
Bosnia and Herzegovina	2,210	2,185	2,185
Botswana	13,718	12,535	11,943
Brazil	520,027	493,213	477,698
British Virgin Islands	4	4	4
Brunei Darussalam	313	288	278
Bulgaria	3,327	3,375	3,625
Burkina Faso	7,154	6,914	6,794
Burundi	289	198	152
Cambodia	12,946	11,541	10,447
Cameroon	24,545	22,345	21,245
Canada	310,134	310,134	310,134
Cape Verde	58	82	84
Cayman Islands	12	12	12
Côte d'Ivoire	10,222	10,328	10,405
Central African Republic	23,203	22,903	22,755
Chad	13,110	12,317	11,921
Channel Islands	1	1	1

Continued

Units: '000 hectares

Country	1990	2000	2005
Chile	15,263	15,834	16,121
China	157,141	177,001	197,290
Colombia	61,439	60,963	60,728
Comoros	12	8	5
Congo Rep. of the	22,726	22,556	22,471
Congo, Dem. Rep. of the	140,531	135,207	133,610
Cook Islands	15	16	16
Costa Rica	2,564	2,376	2,391
Croatia	2,116	2,129	2,135
Cuba	2,058	2,435	2,713
Cyprus	161	173	174
Czech Republic	2,630	2,637	2,648
Denmark	445	486	500
Djibouti	6	6	6
Dominica	50	47	46
Dominican Republic	1,376	1,376	1,376
Ecuador	13,817	11,841	10,853
Egypt	44	59	67
El Salvador	375	324	298
Equatorial Guinea	1,860	1,708	1,632
Eritrea	1,621	1,576	1,554
Estonia	2,163	2,243	2,284
Ethiopia	15,114	13,705	13,000
Faeroe Islands	–	–	–
Falkland Islands	–	–	–
Fiji	979	1,000	1,000
Finland	22,194	22,475	22,500
France	14,538	15,351	15,554
French Guiana	8,091	8,063	8,063
French Polynesia	105	105	105
Gabon	21,927	21,826	21,775
Gambia	442	461	471
Georgia	2,760	2,760	2,760
Germany	10,741	11,076	11,076
Ghana	7,448	6,094	5,517
Gibraltar	–	–	–
Greece	3,299	3,601	3,752
Greenland	–	–	–
Grenada	4	4	4
Guadeloupe	84	81	80
Guam*	26	26	26

Continued

Units: '000 hectares

Country	1990	2000	2005
Guatemala	4,748	4,208	3,938
Guinea	7,408	6,904	6,724
Guinea-Bissau	2,216	2,120	2,072
Guyana*	15,104	15,104	15,104
Haiti	116	109	105
Honduras	7,385	5,430	4,648
Hungary	1,801	1,907	1,976
Iceland	25	38	46
India	63,939	67,554	67,701
Indonesia	116,567	97,852	88,495
Iran, Islamic Rep. of	11,075	11,075	11,075
Iraq	804	818	822
Ireland	441	609	669
Isle of Man	3	3	3
Israel	154	164	171
Italy	8,383	9,447	9,979
Jamaica	345	341	339
Japan	24,950	24,876	24,868
Jordan	83	83	83
Kazakhstan	3,422	3,365	3,337
Kenya	3,708	3,582	3,522
Kiribati	2	2	2
Korea, Dem. People's Rep.	8,201	6,821	6,187
Korea, Rep. of	6,371	6,300	6,265
Kuwait	3	5	6
Kyrgyzstan	836	858	869
Lao People's Dem. Rep.	17,314	16,532	16,142
Latvia	2,775	2,885	2,941
Lebanon*	121	131	136
Lesotho	5	7	8
Liberia	4,058	3,455	3,154
Libyan Arab Jamahiriya	217	217	217
Liechtenstein	6	7	7
Lithuania	1,945	2,020	2,099
Luxembourg	86	87	87
Macedonia, FYR	906	906	906
Madagascar	13,692	13,023	12,838
Malawi	3,896	3,567	3,402
Malaysia	22,376	21,591	20,890
Maldives	1	1	1
Mali	14,072	13,072	12,572

Continued

Units: '000 hectares

Country	1990	2000	2005
Malta	–	–	–
Martinique	46	46	46
Mauritania	415	317	267
Mauritius	39	38	37
Mexico	69,016	65,540	64,238
Micronesia, Federated States of	63	63	63
Moldova, Rep. of	319	326	329
Monaco	–	–	–
Mongolia	11,492	10,665	10,252
Morocco	4,289	4,328	4,364
Mozambique	20,012	19,512	19,262
Myanmar	39,219	34,554	32,222
Namibia	8,762	8,033	7,661
Nauru	–	–	–
Nepal	4,817	3,900	3,636
Netherlands	345	360	365
Netherlands Antilles	1	1	1
New Caledonia	717	717	717
New Zealand	7,720	8,226	8,309
Nicaragua	6,538	5,539	5,189
Niger	1,945	1,328	1,266
Nigeria	17,234	13,137	11,089
Niue	17	15	14
Northern Mariana Islands	35	34	33
Norway	9,130	9,301	9,387
Oman	2	2	2
Pakistan	2,527	2,116	1,902
Palau	38	40	40
Palestinian Territories*	9	9	9
Panama	4,376	4,307	4,294
Papua New Guinea	31,523	30,132	29,437
Paraguay	21,157	19,368	18,475
Peru	70,156	69,213	68,742
Philippines	10,574	7,949	7,162
Poland	8,881	9,059	9,192
Portugal	3,099	3,583	3,783
Puerto Rico	404	407	408
Qatar	–	–	–
Réunion	87	87	84
Romania	6,371	6,366	6,370
Russian Federation	808,950	809,268	808,790

Continued

Units: '000 hectares

Country	1990	2000	2005
Rwanda	318	344	480
Saint Helena	2	2	2
Saint Christopher and Nevis	5	5	5
St Lucia	17	17	17
Saint Pierre and Miquelon	3	3	3
St Vincent and the Grenadines	9	10	11
Samoa	130	171	171
San Marino	–	–	–
São Tome and Príncipe	27	27	27
Saudi Arabia	2,728	2,728	2,728
Senegal	9,348	8,898	8,673
Serbia and Montenegro	2,559	2,649	2,694
Seychelles	40	40	40
Sierra Leone	3,044	2,851	2,754
Singapore	2	2	2
Slovakia	1,922	1,921	1,929
Slovenia	1,188	1,239	1,264
Solomon Islands	2,768	2,371	2,172
Somalia	8,282	7,515	7,131
South Africa	9,203	9,203	9,203
Spain	13,479	16,436	17,915
Sri Lanka	2,350	2,082	1,933
Sudan	76,381	70,491	67,546
Suriname	14,776	14,776	14,776
Swaziland	472	518	541
Sweden	27,367	27,474	27,528
Switzerland	1,155	1,199	1,221
Syrian Arab Republic	372	432	461
Tajikistan	408	410	410
Tanzania	41,441	37,318	35,257
Thailand	15,965	14,814	14,520
Timor-Leste	966	854	798
Togo	685	486	386
Tonga	4	4	4
Trinidad and Tobago	235	228	226
Tunisia	643	959	1,056
Turkey	9,680	10,052	10,175
Turkmenistan	4,127	4,127	4,127
Turks and Caicos Islands	34	34	34
Uganda	4,924	4,059	3,627
Ukraine	9,274	9,510	9,575

Continued

Units: '000 hectares

Country	1990	2000	2005
United Arab Emirates	245	310	312
United Kingdom	2,611	2,793	2,845
USA	298,648	302,294	303,089
Uruguay	905	1,409	1,506
Uzbekistan	3,045	3,212	3,295
Vanuatu	440	440	440
Venezuela	52,026	49,151	47,713
Viet Nam	9,363	11,725	12,931
Virgin Islands	12	10	10
Western Sahara	1,011	1,011	1,011
Yemen	549	549	549
Zambia	49,124	44,676	42,452
Zimbabwe	22,234	19,105	17,540
World	4,077,291	3,988,610	3,952,025

* Guam, Guyana, Lebanon and Palestinian Territories did not provide an estimate of forest area for 1990. Datapoint was estimated by FAO based on a linear extrapolation of the figures provided for 2000 and 2005.

Source: World Resources Institute. 2007. EarthTrends: Environmental Information. Available at http://earthtrends.wri.org. Washington DC: World Resources Institute.

Food and Agriculture Organization of the United Nations (FAO). 2005. Global Forest Resources Assessment 2005: Progress towards sustainable forest management. FAO Forestry Paper 147. Rome: FAO. Available online at: http://www.fao.org/forestry/foris/webview/forestry2/index.jsp?siteId=101&langId=1.

Adapted from: Food and Agriculture Organization of the United Nations (FAO). 2005. Global Forest Resources Assessment 2005: Progress towards sustainable forest management. FAO Forestry Paper 147. Rome: FAO.

Definition: Total forest area is determined both by the presence of trees and the absence of other predominant land uses. Land spanning more than 0.5 hectares with trees higher than five metres and a canopy cover of more than 10%, or trees able to reach these thresholds, such as areas under reforestation and areas temporarily unstocked but expected to regenerate, are considered forests. Forest area does not include land predominantly under agricultural or urban land use, such as tree stands in fruit plantations and agroforestry systems, or urban parks and gardens. Data are measured in thousand hectares.

Forest area includes: areas with bamboo and palms provided that height and canopy cover criteria are met; forest roads, firebreaks and other small open areas; forest in national parks, nature reserves and other protected areas such as those of specific scientific, historical, cultural or spiritual interest; windbreaks, shelterbelts and corridors of trees with an area of more than 0.5 hectares and width of more than 20 metres; plantations primarily used for forestry or protective purposes, such as rubber-wood plantations and cork oak stands.

Table 11: Ratification of Environmental Treaties
(by Human Development Index rank)

HDI rank		Cartagena Protocol on Biosafety	Framework Convention on Climate Change	Kyoto Protocol to the Framework Convention on Climate Change	Convention on Biological Diversity
High human development					
1	Norway	●	●	●	●
2	Iceland	○	●	●	●
3	Australia		●	○	●
4	Ireland	●	●	●	●
5	Sweden	●	●	●	●
6	Canada	○	●	●	●
7	Japan	●	●	●	●
8	USA		●	○	○
9	Switzerland	●	●	●	●
10	Netherlands	●	●	●	●
11	Finland	●	●	●	●
12	Luxembourg	●	●	●	●
13	Belgium	●	●	●	●
14	Austria	●	●	●	●
15	Denmark	●	●	●	●
16	France	●	●	●	●
17	Italy	●	●	●	●
18	United Kingdom	●	●	●	●
19	Spain	●	●	●	●
20	New Zealand	●	●	●	●
21	Germany	●	●	●	●
22	Hong Kong, China SAR				
23	Israel		●	●	●
24	Greece	●	●	●	●
25	Singapore		●	●	●
26	Korea, Rep. of	○	●	●	●
27	Slovenia	●	●	●	●
28	Portugal	●	●	●	●
29	Cyprus	●	●	●	●
30	Czech Republic	●	●	●	●
31	Barbados	●	●	●	●
32	Malta		●	●	●
33	Kuwait		●	●	●
34	Brunei Darussalam				
35	Hungary	●	●	●	●
36	Argentina	○	●	●	●
37	Poland	●	●	●	●

285

Continued

HDI rank		Cartagena Protocol on Biosafety	Framework Convention on Climate Change	Kyoto Protocol to the Framework Convention on Climate Change	Convention on Biological Diversity
38	Chile	○	●	●	●
39	Bahrain		●	●	●
40	Estonia	●	●	●	●
41	Lithuania	●	●	●	●
42	Slovakia	●	●	●	●
43	Uruguay	○	●	●	●
44	Croatia	●	●	○	●
45	Latvia	●	●	●	●
46	Qatar		●	●	●
47	Seychelles	●	●	●	●
48	Costa Rica	○	●	●	●
49	United Arab Emirates		●	●	●
50	Cuba	●	●	●	●
51	Saint Christopher and Nevis	●	●		●
52	Bahamas	●	●	●	●
53	Mexico	●	●	●	●
54	Bulgaria	●	●	●	●
55	Tonga	●	●		●
56	Oman	●	●	●	●
57	Trinidad and Tobago	●	●	●	●
58	Panama	●	●	●	●
59	Antigua and Barbuda	●	●	●	●
60	Romania	●	●	●	●
61	Malaysia	●	●	●	●
62	Bosnia and Herzegovina		●		●
63	Mauritius	●	●	●	●
Medium human development					
64	Libyan Arab Jamahiriya	●	●		●
65	Russian Federation		●	●	●
66	Macedonia, FYR	●	●	●	●
67	Belarus	●	●	●	●
68	Dominica	●	●	●	●
69	Brazil	●	●	●	●
70	Colombia	●	●	●	●
71	Saint Lucia	●	●	●	●
72	Venezuela	●	●	●	●
73	Albania	●	●	●	●

Continued

HDI rank		Cartagena Protocol on Biosafety	Framework Convention on Climate Change	Kyoto Protocol to the Framework Convention on Climate Change	Convention on Biological Diversity
74	Thailand	●	●	●	●
75	Samoa (Western)	●	●	●	●
76	Saudi Arabia		●	●	●
77	Ukraine	●	●	●	●
78	Lebanon		●		●
79	Kazakhstan		●	○	●
80	Armenia	●	●	●	●
81	China	●	●	●	●
82	Peru	●	●	●	●
83	Ecuador	●	●	●	●
84	Philippines	○	●	●	●
85	Grenada	●	●	●	●
86	Jordan	●	●	●	●
87	Tunisia	●	●	●	●
88	Saint Vincent and the Grenadines	●	●	●	●
89	Suriname		●		●
90	Fiji	●	●	●	●
91	Paraguay	●	●	●	●
92	Turkey	●	●		●
93	Sri Lanka	●	●	●	●
94	Dominican Republic	●	●	●	●
95	Belize	●	●	●	●
96	Iran, Islamic Rep. of	●	●	●	●
97	Georgia		●	●	●
98	Maldives	●	●	●	●
99	Azerbaijan	●	●	●	●
100	Occupied Palestinian Territories				
101	El Salvador	●	●	●	●
102	Algeria	●	●	●	●
103	Guyana		●	●	●
104	Jamaica	○	●	●	●
105	Turkmenistan		●	●	●
106	Cape Verde	●	●	●	●
107	Syrian Arab Republic	●	●	●	●
108	Indonesia	●	●	●	●
109	Viet Nam	●	●	●	●
110	Kyrgyzstan	●	●	●	●
111	Egypt	●	●	●	●
112	Nicaragua	●	●	●	●

Continued

HDI rank		Cartagena Protocol on Biosafety	Framework Convention on Climate Change	Kyoto Protocol to the Framework Convention on Climate Change	Convention on Biological Diversity
113	Uzbekistan		●	●	●
114	Moldova, Rep. of	●	●	●	●
115	Bolivia	●	●	●	●
116	Mongolia	●	●	●	●
117	Honduras	○	●	●	●
118	Guatemala	●	●	●	●
119	Vanuatu		●	●	●
120	Equatorial Guinea		●	●	●
121	South Africa	●	●	●	●
122	Tajikistan	●	●		●
123	Morocco	○	●	●	●
124	Gabon		●		●
125	Namibia	●	●	●	●
126	India	●	●	●	●
127	São Tomé and Príncipe		●		●
128	Solomon Islands	●	●	●	●
129	Cambodia	●	●	●	●
130	Myanmar	○	●	●	●
131	Botswana	●	●	●	●
132	Comoros		●		●
133	Lao People's Dem. Rep.	●	●	●	●
134	Pakistan	○	●	●	●
135	Bhutan	●	●	●	●
136	Ghana	●	●	●	●
137	Bangladesh	●	●	●	●
138	Nepal	○	●	●	●
139	Papua New Guinea	●	●	●	●
140	Congo, Rep. of the	●	●		●
141	Sudan	●	●	●	●
142	Timor-Leste				
143	Madagascar	●	●	●	●
144	Cameroon	●	●	●	●
145	Uganda	●	●	●	●
146	Swaziland	●	●	●	●
Low human development					
147	Togo	●	●	●	●
148	Djibouti	●	●	●	●
149	Lesotho	●	●	●	●

Continued

HDI rank		Cartagena Protocol on Biosafety	Framework Convention on Climate Change	Kyoto Protocol to the Framework Convention on Climate Change	Convention on Biological Diversity
150	Yemen	•	•	•	•
151	Zimbabwe	•	•		•
152	Kenya	•	•	•	•
153	Mauritania	•	•	•	•
154	Haiti	○	•	•	•
155	Gambia	•	•	•	•
156	Senegal	•	•	•	•
157	Eritrea	•	•	•	•
158	Rwanda	•	•	•	•
159	Nigeria	•	•	•	•
160	Guinea	○	•	•	•
161	Angola		•		•
162	Tanzania, U. Rep. of	•	•	•	•
163	Benin	•	•	•	•
164	Côte d'Ivoire		•		•
165	Zambia	•	•	•	•
166	Malawi	○	•	•	•
167	Congo, Dem. Rep. of the	•	•	•	•
168	Mozambique	•	•	•	•
169	Burundi		•	•	•
170	Ethiopia	•	•	•	•
171	Chad	○	•		•
172	Central African Republic	○	•		•
173	Guinea-Bissau		•	•	•
174	Burkina Faso	•	•	•	•
175	Mali	•	•	•	•
176	Sierra Leone		•		•
177	Niger	•	•	•	•

• Ratification, acceptance, approval, accession or succession. ○ Signature

Source: Human Development Report 2006, United Nations Development Programme. UN (United Nations). 2006d. 'Multilateral Treaties Deposited with the Secretary-General.' New York. http://untreaty.un.org. Accessed August 2006. Reproduced with permission of Palgrave Macmillan.

Note: Information is as of 28 August 2006. The Cartagena Protocol on Biosafety was signed in Cartagena in 2000, the United Nations Framework Convention on Climate Change in New York in 1992, the Kyoto Protocol to the United Nations Framework Convention on Climate Change in Kyoto in 1997 and the Convention on Biological Diversity in Rio de Janciro in 1992.